D0858365

The Master Builders

The Master Builders

A HISTORY OF STRUCTURAL

AND ENVIRONMENTAL DESIGN

FROM ANCIENT EGYPT TO

THE NINETEENTH CENTURY

HENRY J. COWAN
Professor of Architectural Science
University of Sydney

A WILEY-INTERSCIENCE PUBLICATION

JOHN WILEY & SONS, New York • London • Sydney • Toronto

Copyright © 1977 by John Wiley & Sons, Inc.

All rights reserved. Published simultaneously in Canada.

No part of this book may be reproduced by any means, nor transmitted, nor translated into a machine language without the written permission of the publisher.

Library of Congress Cataloging in Publication Data

Cowan, Henry J.
  The master builders.

  Bibliography: p.
  1. Building—History. 2. Architecture—History.
I. Title.
TH15.C62   690'.09   77-5125
ISBN 0-471-02740-5

Printed in the United States of America

10 9 8 7 6 5 4 3 2 1

For

Dr. Valerie Havyatt

because she likes history

PREFACE

This is the first of two volumes on the history of building science from ancient times to the present day. It relates the story to the end of the Napoleonic Wars. This was the age of the master builders who erected great buildings by essentially empirical rules. The foundations of the modern theory of building science were laid in the seventeenth century. The companion volume, *Science and Building*, explains how this theory transformed the practice of architecture and building in the nineteenth and twentieth centuries. The two books are, however, separate entities and one can be read without reading the other.

I am indebted to Dr. Valerie Havyatt for a great deal of bibliographical research, to Dr. R. Baumann (Department of History, University of Sydney), to Mr. D. Saunders (Department of Fine Arts, University of Sydney), to Mr. A. Wargon (Consulting Engineer), to Dr. Havyatt and Professor P. R. Smith (both of the Department of Architectural Science, University of Sydney) for many helpful comments and corrections, to Mr. John Dixon for all the photographic work, and to Mrs. Rita Arthurson, Mrs. Hilda Mioche, and Miss Ann Novitski for typing the manuscript.

HENRY J. COWAN

Sydney, Australia
June 1977

A NOTE TO THE GENERAL READER

This book is intended for architects, builders, and engineers, but most of it is readily intelligible to readers who are interested in the subject but know nothing of science or the building industry.

There are two short passages that require a little mathematics, namely a proof by Euclid in Section 2.2 and the mechanics of a Greek temple in Section 3.2. These short passages can be omitted without loss of continuity.

Sections 7.4, 8.2, 8.3, and 8.4 assume some knowledge of mechanics. These are marked with an asterisk in front of the section number (e.g., *7.4). Readers who have no previous knowledge of the subject may prefer to omit them.

H. J. C.

CONTENTS

The Master Builders

CHAPTER ONE

Some

Introductory Remarks

Who shall doubt the secret hid
Under Cheops' pyramid
Was that the contractor did
Cheops out of several millions?

RUDYARD KIPLING

Departmental Ditties and Other Verses 1891

The opening chapter discusses some of the problems encountered in writing this book.

1.1 MODERN BUILDINGS AND ANCIENT BUILDINGS

How does one compare the Egyptian pyramids with a modern block of low-income apartments? This is the problem faced by any author who is endeavoring to write a history of building.

Little has survived from the ancient world and the Middle Ages except religious and sepulchral buildings, yet few modern architects spend much time on the design of churches and even fewer design tombs. The real precursor of modern architecture is therefore the housing used by the ordinary people of ancient and medieval times.

It is possible, although by no means certain, that the primitive housing of the developing countries today is similar. The highlands of New Guinea until quite recently had virtually no contact with the outside, and the methods of home building still used there (Fig. 1.1) may resemble those of the common people of the ancient world. The mud huts of Africa are another possible prototype (Fig. 1.2).

For more complex buildings we have at least two examples: the buried cities of Pompeii and Herculaneum discovered in 1594, rediscovered in 1748, and excavated since 1860 are one; the other is the Inca city of Machu Pichu, which was abandoned intact, lost in the jungle, and rediscovered in 1911. These, however, are rare exceptions to the rule that surviving ancient buildings are mostly religious or sepulchral; surviving residences are generally the palaces of rulers.

Evidently houses or blocks of apartments which relate much more closely to modern architecture than temples and tombs were built for ordinary people in all civilizations. We can gain some information from excavations (Section 3.8) and from Vitruvius' writings (Sections 2.1 and 3.4). Surprisingly enough, we know more about ancient Rome than about the Middle Ages.

The size of the population of that early metropolis is in dispute (Section 3.8), but at its peak it was probably comparable to that of London at the beginning of the nineteenth century, when, with approximately a million people, it had become the most populous city in Europe. Both had similar advantages and disadvantages. There were better employment opportunities, better communal facilities, and greater freedom to live one's own life, but there were also the problems of overcrowding, noise, pollution, and lack of adequate transport.

The fascination of nineteenth-century England with "the grandeur that was Rome" was at least partly due to the lessons to be learned not merely in the governing of a great empire but in more immediately practical matters such as public hygiene. In 1842 Edwin Chadwick (Section 4.3) pointed out that the Romans had a better system of water supply and sewage disposal than contemporary Europe.

1.2 A NOTE ON THE SOURCES OF INFORMATION

Although information on recent events is abundant, relevant writings before the midnineteenth century are fewer. Sometimes they are not easy to understand

1.1

Circular hut made of bamboo tied together with vegetable fibers. The walls are woven mats and the roof is thatch. The hut can be quickly erected by two people and is reasonably comfortable in a hot-humid climate, but it lacks durability.

1.2

Mud huts common in various parts of Africa, particularly in the hot-arid region. These huts differ from the traditional type only by having a corrugated iron roof which is now common practice (see also Fig. 5.8).

because mathematical derivations employed techniques no longer in use and technical terms have changed their meaning. Important buildings have been altered without making the accurate records of their previous condition taken for granted in modern restorations. Restorers, particularly of Gothic buildings, have sometimes used interpretations of the designer's original intention that we now believe to be incorrect (Section 6.6), and it is not always possible to tell what part of the building is original and what part, restored. Before the eighteenth century most people had little interest in historical buildings except as useful artifacts and restoration meant making them habitable for their new occupants.

Going back in time, we can establish the relevant facts with reasonable accuracy as far as the fifteenth century, but before the invention of printing written information on building science is scarce. It was laborious to write a manuscript and paper and parchment were valuable. Many ancient manuscripts have been destroyed to make use of the material on which they were written: for example, the celebrated sketchbook of Villard de Honnecourt (Ref. 1.2 and Section 6.3) is in part a palimpsest; some pages are written on material from which the original writing has been effaced.

The survival of the payrolls of some buildings (Section 6.3) enables us to estimate the time and method of construction, the materials employed, and the size and composition of the workforce. Occasionally we find writings of interest to building science engraved in stone; a carved illustration of a building operation or piece of machinery (Fig. 2.5) is invaluable evidence but regrettably rare.

A few timber structures date from the Middle Ages (Section 5.6). None survives from ancient Rome, but we do know a good deal about them from the writings of several Roman authors (Section 3.4).

We find Roman concrete, brick, and stone buildings scattered over the Middle East and in southern and western Europe. Some, like the Pantheon (Fig. 3.20), are still in use. Some, like the Colosseum (Fig. 4.6), have been used as quarries for building material; the construction is clearly visible, however, and it has been there for successive generations to see. Some, like the Roman Bath in Bath, England (Fig. 3.21), have only recently been excavated, but their design is now readily discernible. Of others we have only inadequate remains, and there are probably many whose existence has been forgotten.

We have no original manuscripts from Greek or Roman times, but the literary scholarship of the late nineteenth and early twentieth centuries has established the text of a number of Greek and Roman books relating to building science (Chapter 2). Vitruvius' *Ten Books on Architecture* are important not merely because of the light they throw on Roman ideas and methods but also because of the influence they had during the Renaissance (Section 7.1).

We have more written information about the building science of Rome than about that of the Middle Ages, partly because medieval Europe produced few books. A good medieval library fitted into a cupboard. Vitruvius, telling us how to design a library for a private house, wrote about a room of some size (Section 4.5). The greater literacy of Rome and the prosperity created by more than two centuries of peace, except at remote frontiers, helped to create a society with architectural problems in some ways remarkably similar to those of the nineteenth century.

It is difficult to obtain much useful information on building science that predates ancient Rome. Many of the techniques described by Vitruvius may have been used by the Greeks and possibly the Egyptians, Phoenicians, and Assyrians, but there is insufficient evidence to form a satisfactory opinion. There is no really useful scientific information on the undoubtedly great civilizations of early India. Although excavations yield plans and other helpful architectural data, they tell us little about the superstructures, apart from the materials employed.

This book barely mentions Japan and China. Modern Japanese building science, which is one of the most sophisticated to be found in any country, owes little to tradition. This is not a criticism of ancient Japanese architecture which has influenced not only modern Japanese architecture but modern architecture throughout the world. It is a statement that Japanese building science is derived from Europe, partly by way of America, rather than from Japanese tradition. English-language abstracts and journals, published in China, convey the impression that Chinese building science is also based on the European tradition. Some scientific ideas (Sections 7.1 and 8.6), however, did come to Europe from China (Ref. 1.1).

One difficulty in interpreting the architecture of the past is the deplorable habit of many conquerors of demolishing prestigious buildings and raising their own most important buildings on the ruins; this not merely destroyed the building but made excavation of the remains more difficult. Presumably this was done partly as a mark of triumph and partly to attract the loyalty attaching to the site.

In recent years the foundations of the Great Pyramid, the most important Aztec religious monument, were found under the cathedral in Mexico City. Justinian's Palace has been excavated under the Blue Mosque in Constantinople (Fig. 5.3), and the remains of a Roman temple are presently being excavated under the Duomo of Florence.

The Pantheon (Section 3.9) would perhaps have survived in any case because its massiveness makes it almost indestructable, but its conversion from a pagan temple to a church guaranteed its preservation. The conversion of the church of S. Sophia (Section 5.2) into a mosque also ensured its survival. The Parthenon (Section 3.2) was less fortunate.

We have benefited from some natural disasters; for example, the volcanic eruption of Mt. Vesuvius, which buried the cities of Pompeii and Herculaneum in A.D. 79 and killed most of the inhabitants, has provided us with much valuable information on the normal life of a Roman city.

1.3   THE SCIENCE OF ARCHITECTURE AND THE ART OF ARCHITECTURE

In any discussion of the history of architectural science it is proper to point out that the terms art and science in relation to architecture have, over the centuries, been almost exactly reversed in their meaning. Today, architectural science generally includes the study of structures and materials, the applications of physics (heat, light, and sound) to architectural design and the hardware for the building services, and, recently, the application of computers to architectural planning and building economics. One aspect normally excluded is architectural aesthetics.

In medieval and Renaissance texts the term *scientia* referred to geometry and the theory of proportions, the only part of architectural practice amenable to mathematical treatment at the time (Sections 6.3 and 7.11). The term *ars* generally referred to craft practices (the "useful arts" of the nineteenth century).

1.4 ABOUT THIS BOOK

The history of building technology can be divided into two distinctive parts. The earlier, during which the technology was mainly empirical, and the later, during which building technology was based increasingly on science.

I have chosen 1815, the date of the Battle of Waterloo, as the end of the first period. The causes of industrialization, which changed the character of the cities of Europe, belong to the preceding century, as do the first iron structures and the beginnings of the mathematical theory of structures. They had, however, only a slight effect on the design and construction of buildings before the end of the Napoleonic Wars. The subsequent period is covered in a separate book (Ref. 1.3).

The history of building technology before 1815 can itself be divided into four parts. The first terminates in A.D. 476 with the fall of the West Roman Empire. It consists of Chapters 2 to 4 and deals mainly with Rome in the first century B.C. and the first and second centuries A.D. We know a great deal about the building technology of the Roman Empire, much less about the Greek contribution, and very little about those of the ancient Middle Eastern civilizations.

The second period, consisting of Chapters 5 and 6, takes us approximately to the year 1500 and concentrates on the unique structure devised by the Gothic master masons. Chapter 6 includes some late Gothic phenomena, such as the construction of the spire of Beauvais Cathedral in the sixteenth century and the structural concepts of nineteenth-century Neo-Gothic architecture.

Chapter 7 deals with the sixteenth and seventeenth centuries. The emphasis is on the construction of masonry domes, and in this respect it steps outside these centuries at both ends. The dome of Florence Cathedral, the most significant structure of the Renaissance, was built in the early fifteenth century, but the historically oriented contributions to the theory of masonry domes belong to the present era.

Chapter 8 covers the eighteenth century which includes the Industrial Revolution, the first iron structures, and the oldest contributions to structural theory still in use today.

CHAPTER TWO

Roman and Greek Books

Relevant to

Building Science

He who can, does.
He who cannot, teaches.

GEORGE BERNARD SHAW

Man and Superman

We now examine briefly the Roman and Greek books that deal with topics related to the subject matter of this book. Among the several Roman authors on building technology Vitruvius is the most significant. We cannot gauge the exact extent of his influence on ancient Rome, but his was the most important book during the Renaissance.

We know of no Greek books on building science proper, but many Greek texts on mathematics and natural science have survived. The Greek contribution to these subjects was not surpassed until the seventeenth century.

2.1 VITRUVIUS AND OTHER ROMAN AUTHORS

We mention Vitruvius many times because he is the only prefifteenth century author of a major treatise on architecture and building construction whose work has survived.

Vitruvius made extensive references to the Greeks and his style strongly suggests that he copied from earlier Greek texts, but he gave few references and we do not know the Greek sources.

We have no authentic manuscripts of the works of any ancient writer written during his lifetime; therefore establishing the authenticity of the text of any ancient work, including the Bible, is a scientific study in its own right.

We have, of course, the evidence of the buildings and the brief inscriptions in stone that have survived. However, surviving manuscripts are invariably copies made centuries later and it is difficult to know how much they might have been altered in the process.

The ancient attitude toward authorship was different from our own. We would regard it as natural to claim credit for our own work. A modern author is occasionally accused of plagiarism, that is, of copying from an obscure source and claiming it as his work. An ancient author was more inclined to do the opposite; he might be tempted to slip his own modern views into an ancient text to get them accepted as part of an established classic. In some cases the additions are obvious and not in any way deliberate attempts to deceive, as, for example, when medieval and Renaissance buildings were introduced into sixteenth-century editions of Vitruvius.

Vitruvius was not the only Roman author to deal with building construction, and the authenticity of his *Ten Books* is supported by other texts. Both Cato and Pliny offered advice on the proper method of building and Frontinus described the construction of aqueducts. Two short books by Palladius and Faventinus (Ref. 3.29), which dealt mainly with building construction, were written at a time when concrete was well established; that is, during the Empire.

Vitruvius' *De Architectura Libri Decem* (Ten Books on Architecture, Ref. 2.3), has acquired a unique position. All the authors of the Renaissance, from Alberti (Ref. 2.4) to Wotton (Ref. 2.5), owe him a debt, and even modern books on design still quote with approval Wotton's sentence: "Well building hath three Conditions. Commoditie, Firmenes, and Delight." (*Of the Elements of Architecture*, Part I). This was modeled on Vitruvius' three conditions.

Thus Vitruvius' writing is comparable in authority to that of Aristotle on natural science, Euclid on geometry, or Ptolemy on astronomy.

Vitruvius, in the preface of Book 1, claimed to have been a military engineer in the service of Julius Caesar. In Book 5, Chapter 1, he stated that he built the basilica for the tribunal at Fano. Frontinus (see Section 4.3) mentioned him as having been employed on the construction of an aqueduct. That is all we know about his career. Morgan (Ref. 2.3) thought that he must have written during the reign of Augustus. Fensterbusch, who made a recent German translation (Ref. 2.1), dated his book more precisely between 27 and 14 B.C. Vitruvius was quoted by Pliny and Frontinus in the first century A.D., by Faventinus in the third century, and by Sidonius Appolinaris in the fifth.

There are fifty-five ancient manuscripts of Vitruvius, of which the oldest is believed to date from the ninth century. In most cases it is not known where the manuscripts were written and there are discrepancies among them (see Section 7.1).

The first Latin edition has no imprint of the date, place, or name of the printer, but it is believed to date from 1487 (Ref. 2.1). Further Latin editions appeared in Florence in 1495, in Venice in 1497, and at least ten were published during the sixteenth century.

Vitruvius is best known in translations. The Italian translations by Cesariano (Milan, 1521) and by Barbaro (Venice, 1556) were particularly influential. The first English translation was published in London in 1730. At present the most commonly used English translation (Ref. 2.3) is that made by Morgan (Oxford, 1914).

Vitruvius' influence is mainly due to the importance attached by the Renaissance to the correct use of the Greek and Roman orders; that is, the exact proportions to be used for the various parts of classical columns, architraves, and other features which form the "scientific" part of Renaissance architecture (see Sections 1.3 and 7.11) and for which mathematics was needed. From our point of view the book is of special interest for the evidence it provides on the Roman approach to building science in its modern sense.

Roman measures varied a little with time and place. The following conversions are derived from the commentary by Fensterbusch (Ref. 2.1, p. 543):

1 cubitus (cubit, the length of a forearm) = 1½ pedes = 443.4 mm = 17.5 in.

1 pes (Roman foot) = 4 palmae = 295.6 mm = 11.6 in. = 0.97 ft

1 palma (palm) = 4 digiti = 73.9 mm = 2.9 in.

1 digitus (digit, the length of a finger) = 18.4 mm = 0.73 in.

2.2 GREEK AND ROMAN SCIENCE AND MATHEMATICS

We can deal only briefly with the history of science, for it is a subject so much larger and more complex than the history of building science that even an adequate outline would take up an entire volume. The reader is referred to Refs. 2.6 to 2.12 at the end of the book.

The Babylonians, Egyptians, and Phoenicians made noteworthy contributions to science, but they were overshadowed by the Greeks. Greek mathematics, in particular, was not surpassed until the seventeenth century.

Thales of Miletus (now Balat in Turkey), who lived in the sixth century B.C., is generally considered the first Greek mathematician. The end of Greek science came with the Arab conquest of Alexandria in A.D. 641.

Pythagoras, the best known of the early mathematicians, was born on the Aegean island of Samos in 569 B.C. and migrated to Sicily in 529 where he died about 500. Apart from the famous theorem of right-angled triangles, he discovered the fixed intervals in the musical scale (which are the basis of the harmonic proportions of architecture) and contributed to the theory of numbers.

About 420 B.C. Athens established itself as the center of Greek science. Plato lived there from 428 to 348 B.C. and about 380 founded a school he called the Academy. Plato is best known as a philosopher, but his contributions to the physical and biological sciences were also important. Through the neo-Platonists of Alexandria he influenced St. Augustine, who was Bishop of Hippo in Roman Africa about A.D. 396 to 430, and Boethius, a Christian philosopher who lived in Rome from 480 to 524. Both were highly esteemed by medieval scholars and Greek ideas were revived through them in Europe before the recovery of the Greek texts (see Section 6.3).

In *Timaeus* Plato argued that all beauty depends on perfect proportions. In *De Musica* Boethius explained that "the ear is affected by sounds in quite the same way as the eye is by optical impressions" (Ref. 6.5, p. 33). Possibly in the Middle Ages, but certainly in the Renaissance, this gave rise to the system of harmonic proportions (see Sections 6.3 and 7.11).

Aristotle was born in Macedonia in 384 B.C. His father was physician to Philip of Macedonia. Aristotle was a member of Plato's Academy from 367 to 347, in which year he was appointed tutor to Alexander. When Alexander ascended the throne in 336, Aristotle returned to Athens and founded his own school, the Lyceum. He retired to Chalcis in 321 B.C. and died there a year later. Through the works of the two great Arab philosophers Avicenna (who died in Hamadan in 1037) and Averroes (who died in Cordoba in 1198) and St. Thomas Aquinas (1225–1274) he became *the* philosopher of the late Middle Ages and the Renaissance.

Aristotle's influence on building science was not entirely helpful. He did not invent the elements, but it was his doctrine that dominated chemistry until the seventeenth century. Aristotle stated that all earthly substances are composed of four basic *elements,* termed earth, water, air, and fire, which are ideal components with the general properties of earth, water, air, and fire, not actual earth, and so on. We shall see in Sections 3.5 and 4.1 that Vitruvius thought it necessary to support his perfectly rational observations of the durability of stone and timber and of the behavior of pozzolana by a completely erroneous explanation in terms of the four elements. This practice of explaining chemical changes was resumed in the late Middle Ages and continued throughout the Renaissance. It evidently hampered a rational consideration of the behavior of materials, but most Renaissance scholars considered a show of conformity with Aristotelian natural philosophy to be essential.

In 332 B.C. Alexander conquered Egypt and in 331 laid out the city of Alexandria. He died in 323 and the empire was divided between his generals. Ptolemy, who became king of Egypt, chose Alexandria for his capital. In 306 he decided to create a university, which he called the Museum. Buildings were started and scholars recruited, particularly from Athens. The Museum was opened about 300 and became the center of science and mathematics for almost a thousand years. The best known works of Greek science date from the period 300 to 30 B.C., when, following Cleopatra's suicide, Alexandria became the capital of Roman Egypt. Euclid, Hipparchus, and Hero (see Section 2.3) worked in Alexandria during that time and Archimedes (see Section 2.5) was trained at the Museum.

The best known book on Greek mathematics is *The Elements,* a collection of proofs compiled in Alexandria by the Greek mathematician Euclid who lived about 300 B.C. The geometry taught in present-day high schools is still based on Euclid's *Elements.* The modern editions of Euclid are derived in most part from a Greek manuscript by Theon in the fourth century of the Christian era, that is, 600 to 700 years after Euclid, which was translated into Arabic. The first Latin text was made about A.D. 1120 by Abelard of Bath, who obtained the Arabic version in Spain to which he had gone disguised as a Muslim student.

Let us, as an example of the kind of argument employed by Euclid, examine the last proposition of Book 10, in which Euclid proved that the side and the diagonal of a square are incommensurable (Ref. 2.11, p. 60); that is, if one is an integral number, the other cannot be an integral number.

Let us assume that the side of the square a and its diagonal b are in a commensurable ratio; that is, a and b are two integers. Let us reduce this ratio to the lowest terms so that a and b have no common divisor other than unity; that is, they are prime to one another. Because the diagonal is the hypotenuse of a right-angled triangle, $b^2 = 2a^2$ (Euclid, I, 47). Therefore b^2 is an even number and b is an even number. Because a is prime to b, a must be an odd number. Again, because it has been shown that b is an even number, we can put $b = 2n$; therefore $(2n)^2 = 2a^2$; therefore $a^2 = 2n^2$; therefore a^2 is an even number and a is an even number. Thus the same number a must be both odd and even, which is absurd. Therefore the assumption is incorrect and a and b must be incommensurable.

It follows that a proportional system based on the side of a square and its diagonal cannot be translated accurately into arithmetic (see Section 7.11). Euclid also derived the construction of the Golden Section, but he used it for the construction of the regular pentagon. Its use in aesthetics dates only from the nineteenth century (see Ref. 1.3, Section 10.8).

Trigonometry was devised by Hipparchus, the most eminent of Greek astronomers, who worked between 161 and 127 B.C. He spent some time in Alexandria but made most of his observations in Rhodes.

The Museum declined after the Roman conquest of Egypt in 30 B.C. but it continued to function for another 600 years. The most famous scholar of that second period was Ptolemy, who died in A.D. 168. His *Megiste Mathematike Syntaxis* (Great mathematical compendium) was translated into Arabic, known from a corruption of the Greek title as the Almagest, and remained the standard work on astronomy, navigation, and surveying for more than a thousand years. It contained the equations of the motions of the heavenly bodies on the assumption that the earth is the fixed center of the universe. It also contained tables of trigonometric functions, described instruments for making stellar observations, and gave trigonometric equations for computing the results. The astronomical theory was rendered obsolete by the work of Kepler and Galileo, but the usefulness of the data has not been affected, for navigation and surveying are still for the convenience of computation based on a hypothetical universe in which the earth is fixed and the stars move around it.

A modern architect who has mastered as much geometry and trigonometry as was known to a mathematician in Alexandria in the first century B.C. knows more than he requires of these two branches of mathematics. He would also know more than appears to have been available to Vitruvius.

The *Ten Books on Architecture* contain a lot of numerical data and simple ratios to determine the correct proportional relationships of the orders. The calculations could easily have been done with an abacus which the Romans used for their arithmetic.

Vitruvius gave only a few constructions involving geometry and for the two most complex he referred us to diagrams. One dealt with the *entasis* of columns, which Vitruvius described as

proportionate enlargements made in the thickness of columns on account of the different heights to which the eye has to climb. For the eye is always in search of beauty and if we do not gratify its desire for pleasure by a proportionate enlargement in these measures, and thus make compensation for the ocular deception, a clumsy and awkward appearance will be presented to the beholder. With regard to the enlargement made at the middle of the columns, which among the Greeks is called entasis, at the end of the book a figure and calculations will be subjoined showing how an agreeable and appropriate effect may be produced by it. (Ref. 2.3, Book 3, Chapter 3, p. 86)

We no longer have the original diagram and calculations—only the substitutions of later centuries which get more complex in the later editions.

The other diagram dealt with the construction of the volutes for the scroll-like capital of the Ionic order. Vitruvius in Book 3, Chapter 5 (Ref. 2.3, p. 92), described the construction of the spirals, but there is insufficient detail in the text to perform the construction. Vitruvius referred to a diagram that is missing. The Greek and Roman orders used in the Renaissance were mostly based on *L'Architettura,* published between 1537 and 1551 by Sebastiano Serlio who filled in the details, and not directly on Vitruvius.

The complexity of the geometric constructions in architectural design increased steadily from the sixteenth century on. Guarini (Ref. 2.13), in a book published in 1683, was already printing geometry that would baffle many modern architects.

Roman geometry lagged behind that of Greece. Rouse Ball (Ref. 2.11, p. 113) stated that the mathematical schools in Rome confined their teaching to the art of calculation with the aid of an abacus, to a few practical rules of geometry, and to a little trigonometry. Those who wanted to know more went to study in Alexandria. The fact that Vitruvius employed only a small amount of the geometry known in his time does not prove that Greek architects might not have used more.

Since the eighteenth century a number of theories have been proposed in regard to the geometry employed in the design of classical Greek architecture and most have been proved by applying them to the measured dimensions of famous Greek buildings, particularly the Parthenon. Different theories often fit the same building within a reasonable margin of error. However, if one of the theories that explain the design of a particular building is correct, the other theories that fit the same building must be incorrect and it is possible that all the modern explanations are incorrect.

The Parthenon was built about 440 B.C. and the Temple of Aphaia (see Section 3.2) was probably under construction from 447 to 438, a century after Thales and Pythagoras; the Temple of Artemis (Diana) of Ephesus (see Section 3.1) was built about 330 B.C., half a century after Eudoxus discovered the Golden Section and about the same time as Euclid was born.

Because most of the geometry in the *Elements* is believed to be a collection of earlier work, it is conceivable that the architects of the Greek masterpieces used elaborate geometric constructions. It is also conceivable that Greek geometers did not see the relevance of their philosophical studies to the practical business of building and vice versa. This lack of correlation between theory and practice has bedeviled building science throughout the ages and is not unknown today.

It seems likely that Vitruvius, who wrote with great deference of Greek achievements, explained their procedures quite correctly. Actually, the Greek orders can be constructed with simple proportional relationships, except for a few details such as entasis and the volutes of the Ionic columns, which could have been drawn with a simple construction such as a spiral.

Greek mathematics nevertheless is relevant to our discussion because it became important two thousand years later, after the Renaissance.

2.3 GREEK PHYSICS

We discuss the acoustics of Greek auditoria, as related by Vitruvius, in Section 4.4. Following his description of the design of the theaters, Vitruvius explained the theory of harmonics of Aristoxenus of Tarantum, a pupil of Aristotle who in the fourth century B.C. wrote a treatise on the *Elements of Harmonics*. This was mainly concerned with the Greek theories of musical intervals and the construction of the diatonic, chromatic, and enharmonic scales. Its value to the science of acoustics is limited.

The musical scales become important to architectural design if one believes that to make buildings beautiful "we must take our proportions from the musicians, who are the greatest masters of this sort of numbers." This view was expressed in 1485 by Alberti (Ref. 2.4, p. 196) and widely held from the Renaissance to the nineteenth century (see Section 7.11). There is no evidence that the Greeks used harmonic theories of proportion.

The Greeks made interesting, but not really useful, contributions to pneumatics, the term given by Greek and Renaissance scientists to the mechanics of air and steam. The most important book on the subject was the *Pneumatika* of Hero of Alexandria, written between the first century B.C. and the first century A.D. This has also come to us in Arab translations, which in turn were translated into Latin. The first English translation by B. Woodcroft and J. G. Greenwood appeared in London in 1851. The original diagrams have been lost and those that follow are reconstructions from the British edition.

Figure 2.1 shows a coin-in-the-slot machine that utilized a lever and piston. Figure 2.2 shows a remarkably ingenious machine operated by steam for opening doors; perhaps it was intended to impress a superstitious congregation or it may just have been a toy to amuse the king of Egypt. Figure 2.3 shows a steam turbine. Burstall constructed a working model (Ref. 2.14, p. 78) that was capable of producing a small amount of power. The concept is quite similar to that of the reaction steam turbine produced by Sir Charles Parsons in 1884.

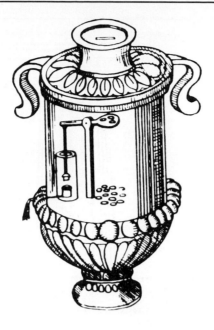

2.1

A coin-in-the-slot machine for dispensing holy water, described by Hero of Alexandria. It contains a lever mechanism not unlike that of a modern vending machine which operates a piston to release a measured quantity of water.

Why did these ingenious machines not produce an industrial revolution in Hellenistic Alexandria or subsequently in Imperial Rome? One reason was the reliance of both Greeks and Romans on slave labor, which made the concept of mechanical power seem superfluous; there was no shortage of human energy. Another was the division in Greek society between the gentleman-philosopher and the builder or mechanic who performed the physical labor. It was proper for a Greek gentleman to play games and to fight and physical prowess was much admired, but it was not proper for him to work with his hands. Roman officers and administrators occasionally engaged in manual work, but the dignity of labor is essentially a Jewish and Christian concept.

Consequently Greek scientists devised ingenious toys but not useful machines, except for waging war. Machines made more progress in the scientifically backward European Middle Ages (see Section 5.4) than they had done in Alexandria's Golden Age of Science.

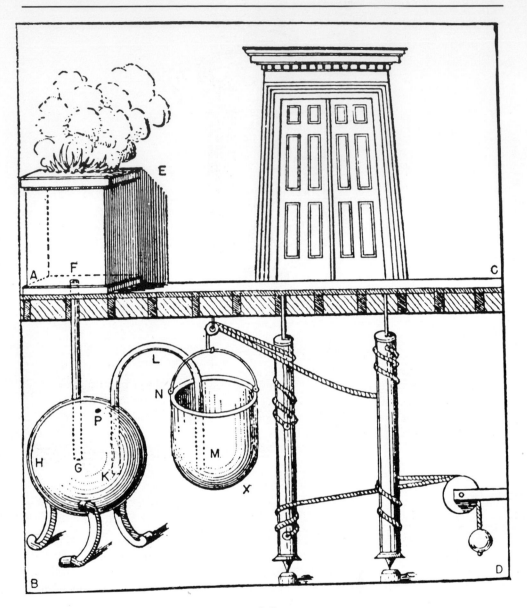

2.2

A mechanism for opening the doors of the temple when a fire was lit on the altar, described by Hero of Alexandria in the *Pneumatika*, Book 1, Chapter 38:

The construction of a small temple such that, on lighting a fire, the doors shall open spontaneously and shut again when the fire is extinguished. Let the proposed temple stand on a pedestal ABCD, on which lies a small altar, EA.

Through the altar insert a tube, FG, of which the mouth F is within the altar, and the mouth G is contained in a globe, H, reaching nearly to its centre: the tube must be soldered into the globe, in which a bent siphon, KLM, is placed. Let the hinges of the door be extended downwards and turn freely on pivots in the base ABCD; and from the hinges let two chains, running into one, be attached, by means of a pulley, to a hollow vessel, NX, which is suspended; while other chains, wound upon the hinges in an opposite direction to the former, and running into one, are attached, by means of a pulley to a leaden weight, on the descent of which the doors will be shut. Let the outer leg of the siphon KLM lead into the suspended vessel; and through a hole, P, which must be carefully closed afterwards, pour water in the globe enough to fill one half of it. It will be found that when the fire has grown hot, the air in the altar becoming heated expands into a larger space; and passing through the tube FG into the globe, it will drive out the liquid contained there through the siphon KLM into the suspended vessel, which, descending with its weight, will tighten the chains and open the doors (Ref. 2.9, p. 328).

2.3

A steam turbine described by Hero of Alexandria in the *Pneumatika*, Book 2, Chapter 11:

Place a cauldron over a fire: a ball shall revolve on a pivot. A fire is lighted under a cauldron, AB, containing water, and covered at the mouth by a lid CD: with this the bent tube EFG communicates, the extremity of the tube being fitted into a hollow ball, HK. Opposite to the extremity G place a pivot, LM, resting on a lid CD: and let the ball contain two bent pipes, communicating with it at the opposite extremities of a diameter, and bent in opposite directions, the bends being at right angles and across the lines FG, LM. As the cauldron gets hot it will be found that the steam, entering the ball thorugh EFG, passes out through the bent tubes toward the lid, and causes the ball to revolve (Ref. 2.9, p. 254–255).

This is a perfectly adequate model of a reaction steam turbine: the passage of the steam is accompanied by a reaction that causes the sphere to rotate in the opposite direction.

2.4 HOISTING GREAT WEIGHTS

The great skill shown in the handling of heavy weights was mainly derived from machines of war. Even in the ancient world building profited from advances in military engineering.

We have many accounts of the construction of military equipment, particularly the siege machines built by the Romans; there is a chapter on this subject in Vitruvius' *Ten Books on Architecture*. These machines involved the handling of extremely heavy weights, which is, of course, of the greatest concern in building construction.

The Egyptians cut huge obelisks from hard granite. The initial incision was probably made by cuts with a chisel, followed by pounding with a ball of diorite to fracture the stone. This may have been followed by the use of wedges, which were moistened to expand them. Unwanted granite was sometimes removed by cracking the stone with fire (Refs. 2.2 and 3.16). The obelisks were moved to the Nile, transported downstream, and moved to their final destinations. They were probably placed by digging two holes; one was filled with sand which was then allowed to run into the adjacent hole to make the obelisk stand up in the first hole as it emptied. This method is portrayed on an obelisk of Rameses II (thirteenth century B.C.), now standing in the Place de la Concorde in Paris.

The Romans moved these obelisks by ship, transporting them across possibly rough seas to Puteoli or Ostia, then overland to Rome, where they were erected.

The achievement of the Renaissance was much more modest. When Fontana in 1586 moved an obelisk from the back of the Basilica of S. Pietro to its present position in front of it, it was acclaimed as a great feat (Section 7.5).

The skill of the Egyptians and Romans, and to a lesser extent the Greeks, in moving huge weights was therefore remarkable, but we do not know exactly how they did it. We have the writings of Archimedes, Hero (Fig. 2.4), and Pappus on the theory of handling weights and we have descriptions of the war machines. One of the few clear illustrations of a Roman crane is shown in Fig. 2.5. Evidently the concept, if not quite so sophisticated as Hero's compound lifting tackle, is the same. It is also a more efficient piece of equipment than any illustrated by Alberti in the fifteenth century (Ref. 2.4 and Section 7.5).

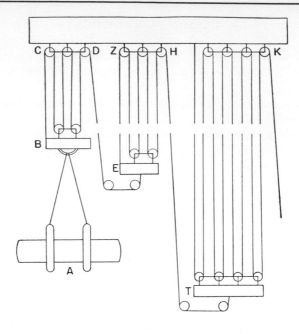

2.4

Compound lifting tackle described by Hero of Alexandria in *Mechanics*, Book 2, Chapter 23:

Let it now be required to move the same weight with the same force by a machine known as the block and tackle system of pulleys. Let A represent the weight, B the point from which it is to be lifted, C the point directly above, that is, the fixed point of support to which we are to lift the weight. Suppose the block and tackle has, let us say, five pulleys, and that the pulley by which the weight is drawn originally is at point D. The force at D must then, in order to balance the 1000 talents, be 200 talents. But the force available is only 5 talents. Therefore we draw a cord from pulley D to block and tackle at a point E. Let there be a fixed point of support at Z directly above E. Let there be at this fixed point of support and at E, in its vicinity, five pulleys, with the pulley at which the force is applied at H. In that case the force at H must be a force of 40 talents. Now if, again, we draw the end of the cord at H to still another block and tackle at T, with fixed point of support at K, at which the pull is exerted, it follows that since 40 talents are 8 times 5 talents, the block and tackle must have 8 pulleys for a force of 5 talents at K to balance the 1000 talents. However, for the force at K to over-balance the weight, the number of pulleys must be more than 8. In that case the force will over-balance the weight (Ref. 2.9 pp. 232–233).

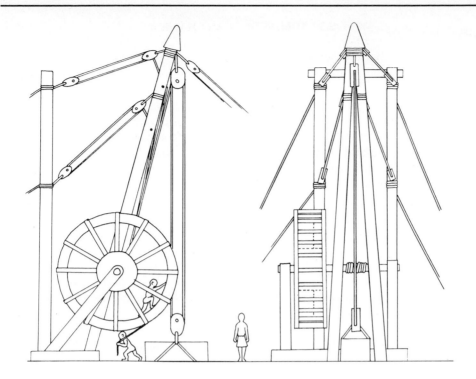

2.5

Roman crane in operation. The left-hand drawing is a line diagram
copied accurately from a relief on a panel in Rome's Lateran Museum;
the right-hand portion is a modern reconstruction of the end-on view,
(Ref. 2.15, p. 189).

2.5 THEORETICAL MECHANICS

Archimedes was born in Syracuse in Sicily in 287 B.C. He attended the Museum in Alexandria and then returned to Syracuse. Like Plato, he expressed the belief that it was undesirable for a philosopher to apply the results of science to any practical use, but he did in fact invent several eminently useful machines. The best known of these is the Archimedian screw, described by Vitruvius in Book 10, Chapter 6 (Ref. 2.3, p. 295), and used in ancient times to remove water from the holds of ships and to drain the fields in Egypt after their annual inundation by the Nile. Archimedes also devised a number of catapults which for some considerable time held the Roman fleet besieging Syracuse in 212 B.C. He was killed when the Romans captured the city in spite of the instructions given by the Roman proconsul Marcellus that his life and house be spared.

In his *Mechanics* Archimedes derived the center of gravity of simple shapes and the equilibrium of levers. Discussing the effect of two weights on a weightless bar, he proved that weights that balance at equal distances are equal and that unequal weights balance at unequal distances, the greater weight being at the lesser distance. He then derived the distances at which these unequal weights are in equilibrium. This was the basis of statics until Stevinus' treatise of 1586, which included the parallelogram of forces (see Section 7.4). Archimedes' lever principle is the oldest contribution to modern structural mechanics and the only one derived from the ancient Greeks and Romans.

CHAPTER THREE

Structure

in the Ancient World

There was once a sculptor named Phidias
Whose manners in art were invidious;
He carved Aphrodite
Without any nightie
Which startled the ultrafastidious.

ANON.

Next we examine Egyptian, Greek, and Roman architecture in the light of modern structural theory. The Romans built the biggest dome before the nineteenth century, but we do not know how they arrived at its dimensions. We are well informed, however, about the Roman methods of laying concrete, masonry, and brickwork and about their fire-fighting services.

3.1 ANCIENT EGYPT, ANCIENT MESOPOTAMIA, AND OTHER PRIMARY CIVILIZATIONS

The ancient world, like the modern, was greatly impressed by sheer size. The Seven Wonders of the World, listed by Antipatros in the first century B.C., and frequently quoted by later Greek and Roman authors, were the following:

1. The Pyramids of Egypt, the only one surviving.
2. The Hanging Gardens of Babylon, possibly a ziggurat with arched terraces carrying roof gardens.
3. The Temple of Artemis (Diana) of Ephesus in Asia Minor, whose main distinction was its great length of 119 m (391 ft) and its use of 120 columns; some of these columns were reused in S. Sophia (see Section 5.2).
4. The statue of Zeus by Phidias at Olympia.
5. The tomb of King Mausolos (the "Mausoleum") at Halikarnassos, in Asia Minor.
6. The Kolossos at the entrance of the harbor of the island of Rhodes (Section 3.2).
7. The lighthouse (Pharos) of Alexandria.

Except for the Kolossos which, if the ancient accounts are correct, was a daring structure, these wonders were remarkable mainly for the mass of material employed.

The monuments that survive from early civilizations frequently consist of piles of stone, like the Egyptian pyramids (Fig. 3.1), piles of mudbrick (e.g., the ziggurats of Mesopotamia), or a facing of carved stone on a natural or artificial hill, as in the Mexican pyramids and in many temples and shrines of Southeast Asia (Fig. 3.2). These structures required a great feat of organization, ingenuity in the handling of materials, and much hard work, particularly in civilizations like those of Mexico and Peru which had only tools of stone (Fig. 3.3).

Massive construction is encouraged by a severe building code. The oldest surviving code is engraved on a column that is now in the Louvre in Paris. It was composed in the reign of Hammurabi, king of Babylon in the eighteenth century B.C. One of the inscriptions lays down: "If a builder has built a house for a man and his work is not strong, and if the house he has built falls in and kills the householder, that builder shall be slain." Evidently this discouraged daring structures like the Gothic cathedrals (see Chapter 6). Several Roman building regulations have survived (Section 3.8); they also tended to emphasize safety rather than economy of material. Because there was no method of determining structural sizes, the ratios of span to depth remained low until the twelfth century A.D.

The problem of span became a main preoccupation of architects from the days of ancient Rome to the early years of the present century; but spans of appreciable size

(a)

(b)

Structure in the Ancient World

3.1

The pyramid of Khufu (a), built about 2700 B.C.* at Gizeh, near Cairo. Its original height of 147 m (482 ft) is now reduced to 137 m. Most of the limestone casing has fallen off and the limestone masonry below has deteriorated (b). The height of this pyramid was not surpassed by any Gothic cathedral (except Beauvais whose tower collapsed; Section 6.5) nor by any Renaissance dome. The spire of Ulm Cathedral, built in the nineteenth century, is slightly taller (159 m). The first multistory structure to exceed it in height was the Singer Building in New York (206 m or 675 ft), erected in 1907.

* Nineteenth century historians assigned a much greater age to the Egyptian monuments. Even in 1905 Banister Fletcher (Ref. 3.20) gave the date of the commencement of the construction of Khufu's pyramid as 3733 B.C. There is, however, insufficient evidence for so precise a date and recent writers have set the time of construction much later. The 1965 edition of the *Encyclopaedia Britannica* (article *Pyramids*) gives 2700 B.C.

3.2

Temple in Bali, Indonesia

3.3a

American Pre-Columbian Architecture

Detail of the temple of Quetzalcoatl at Teotihuacan, near Mexico City, built by the Toltecs about 2000 years ago. The hard igneous rock was shaped with stone tools.

3.3b

American Pre-Columbian Architecture

Squared masonry from Cuzco, the capital city of the Incas in the Peruvian highlands. The stones were carefully fitted by prolonged rubbing with sand; the Inca civilization did not have metal tools. The joints are so tight that it is impossible to push a razor blade between the stones. Mortar was not used.

in a durable material developed much more slowly than the erection of huge mounds of stone which even today are most impressive. We do not know whether this absence of large spans was due to the warm climate in which the early civilizations developed which made outdoor living feasible, to a lack of interest of the early religions in interior spaces, or to their inability to construct them. Nor do we know whether large spans might have been constructed from reeds or timber. These materials were probably much more plentiful in the ancient world (see also Section 3.4) and presumably were used extensively (Ref. 3.1). Unlike masonry, timber has good tensile strength and timber beams can therefore span farther than stone beams. However, timber and reeds are attacked by fungi and insects and they burn. All ancient timber buildings have been destroyed. Fires were started by lightning and by domestic accidents and fire was a major weapon of war. There was no really

3.3c

American Pre-Columbian Architecture

Irregular Inca masonry, fitted with equal precision, from the wall of a
fortress outside the city. The masonry at Machu Picchu, a provincial city,
is similar but less perfectly finished.

effective means of extinguishing a fire once it had taken hold. As a result, permanent
structures had to be roofed in masonry. Alberti (Ref. 2.4, p. 150) in the fifteenth
century put the case thus:

I am entirely for having the roofs of temples arched, as well because it gives them greater
dignity, as because it makes them more durable. And indeed I know not how it happens that
we shall hardly meet any temple whatsoever that has not fallen into the calamity of fire. . . .
Caesar owned that Alexandria escaped being burned, when he himself took it, because its
roofs were vaulted.

3.4

Gateway in Mycaene, Greece. This structure of the pre-Hellenic Mycaenean civilization features a huge stone beam laid without mortar.

Concrete (which is found in ancient buildings in several countries), natural stone, and brick have generally good durability and high compressive strength but their resistance to tension is poor.

The deficient tensile strength of masonry is responsible for the limitations in span of all permanent structures before the eighteenth century and it has largely determined their structural form.

The simplest form of structure consists of a combination of beams and columns (Fig. 3.4). The horizontal members are subject to bending which induces tensile stresses on the lower face and severely limits the span of unreinforced masonry beams (but see Fig. 3.17 on the arching effect of restrained beams).

Appreciable spans can be achieved in unreinforced masonry only by arches and vaulting.* It seems likely that true arches and vaults developed from the use of *corbels* (Figs. 3.5 and 3.6a), which are stones laid in horizontal courses, extending a

* See Glossary for definitions.

Structure in the Ancient World

3.5

Corbeled vault over tomb in Mycenae.

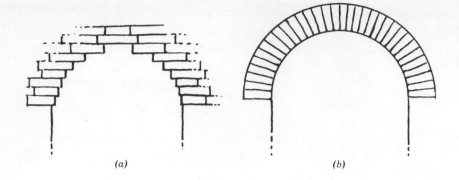

3.6

(a) Corbeled arch—the joints are horizontal.
(b) True arch—the joints are at right angles to the line of thrust.

short distance beyond a wall, each layer a little more than the one below, until the gap is closed. The projecting stones are cantilevers of short span so that the corbels above the opening are subject to tension. This form of construction is best known from the subterranean tombs (Fig. 3.5) belonging to the pre-Hellenic Minoan civilization of Mycenae, which date from about 1200 B.C. (Ref. 3.17, p. 35).

The method is older, however; corbeling has been found in the Bent Pyramid, which was built during the Fourth Egyptian Dynasty (Ref. 3.18, p. 308) about 2900 B.C., and in more recent tombs in both Egypt and Mesopotamia.

The strength of the arch is increased (see Section 3.9) if the joints are aligned at right angles to the lines of thrust (Figs. 3.6b and 3.7a). This is called a *true* arch in which the stones are entirely in compression (see Section 7.3). It is much stronger, but its construction creates additional problems, for we can no longer place the stones in horizontal layers but need formwork to support the sloping joints during erection.

True arches that may date from 2500 B.C. (Ref. 3.18, pp. 512–517) have been found in Egypt, but arches, vaults, and domes never played an important part in Egyptian, Sumerian, Babylonian, Assyrian, or Greek architecture; they were used only for buildings of minor significance (Fig. 3.7b).

Many of the surviving Egyptian monuments are remarkable for their sheer size (Fig. 3.8). The strength of the material was rarely fully utilized (Figs. 3.9 and 3.10) except in roof slabs, and in most structures the horizontal spans are quite small in relation to the materials employed (Refs. 3.18 to 3.21). The many examples of depths that exceed the span suggest a lack of concern for structural economy.

The use of the straight line in Egyptian and, subsequently, in Greek architecture may have been based on aesthetic concepts, as Giedion explained (Ref. 3.18). It may also have been a survival of techniques developed for timber; the mortuary temple of Khafra (Fig. 3.9) is reminiscent of an assembly of timber balks, and even the giant Temple of Amon at Karnak looks like an oversized timber structure (Fig. 3.10).

3.7a

True arch from the Roman theater of Herodotes Atticus in Athens (see also Fig. 4.5).

3.7b

Method of constructing a brick vault without centering used in the Ramesseum (the mortuary temple of Rameses II in Thebes, dating from the thirteenth century B.C.), according to Engelbach (Ref. 3.16, p. 182). The same method is still used by Nubian bricklayers today (Ref. 3.1).

A similar development in the Hindu architecture of India started from rock-cut temples with columns shaped to resemble timber (Ref. 3.14). Later Hindu temples were elaborately carved in a manner that would have been appropriate to timber but was laborious in hard stone (Fig. 3.11).

We can learn much about construction techniques from unfinished work; an example is the incomplete obelisk that still rests in its quarry in Aswan (Section 2.4). In the Temple of Karnak there is an unfinished pylon (Fig. 3.12) and a ramp of mud brick used during its construction is still in position. It is likely that the ancient Egyptians relied on ramps, subsequently removed, rather than scaffolding because of the heavy weight of the individual pieces of stone raised in their temples.

3.8

The temple of Rameses II cut into the solid rock at Abu Simbel, in Nubia, about 1250 B.C. The four statues of Rameses II are about 20 m (66 ft) high. The statue at the center which barely reaches to Rameses' knee is of his favorite queen. Because of the rising waters of the Aswan dam, this temple was cut up in the 1960s, reerected on higher ground, and backed by a reinforced concrete dome which is covered by an artificial mountain. The cuts have been made with such skill that they are invisible in the temple itself.

3.9

Mortuary Temple of Khafra at Gizeh, near Cairo, built about 2600 B.C. This undecorated structure with its square columns and beams is typical of the architecture of the Old Kingdom. Note the cut out in the middle stone of the third row in the back wall to fit the corner stone in the second row. The roof was covered with large stone slabs, some of which are still in position.

Egyptian masonry, unlike that of Greece, was commonly laid in a mortar consisting of gypsum with an admixture of sand and crushed limestone (Ref. 3.16, p. 79).

3.10

Roof beams and columns with lotus-flower capitals from the hypostyle hall of the Temple of Amon at Karnak, Upper Egypt, built about 1250 B.C.

The hall is about 24 m (80 ft) high and has clerestory windows, once covered by stone grills, whose structure is visible in the bottom right-hand corner. It is comparable both in length and height to a Gothic cathedral (Section 6.2). The columns are about 21 m (70 ft) high and their diameter at the base is 3.5 m (11.6 ft); the span of the beams is small.

3.11

Part of a column in the Temple of Vishnu in Kancheepuram near Madras in southern India. It was cut from a hard igneous rock.

3.12

The unfinished pylon at the Temple at Karnak, with the construction ramp of mudbrick still partly in position. The lower stones are not dressed. Egyptian masons, unlike those of Greece, normally finished their stonework in its final location as the construction ramps were removed.

3.2 GREEK MASONRY STRUCTURES

The Hellenic-speaking people came to Greece between 2000 and 1500 B.C. and gradually conquered the inhabitants known to us in the Minoan and Mycenaean civilizations (Fig. 3.4 and 3.5). The earliest significant Hellenic buildings date from 700 B.C., and the most famous Greek temples were all built in a comparatively short period, between the defeat of the Persians in 480 B.C. and the death of Alexander of Macedon in 323. The Romans occupied Greece in the second century B.C.; it then became a protectorate of the Roman Republic.

The Greeks colonized the Aegean islands, Asia Minor, and southern Italy. Some of the best Doric temples are in Sicily and southern Italy and some of the best Ionic temples, in Asia Minor.

In the fifth century B.C. Athens became the main center of Greek science and architecture. The temple of Athena Parthenos (the virgin Athena) on the Acropolis (citadel) of Athens is widely regarded as the most important Greek building (Fig. 3.13).

Most theories of aesthetic proportion have been applied to the Parthenon (see Section 2.2) but it is possible that its dimensions were based on practical considerations. Carpenter (Ref. 3.22) has argued that the Parthenon was started by the architect Kallicrates during the reign of Kimon. When Pericles toppled Kimon from power, he replaced Kallicrates with Iktinos in 448 B.C. and ordered the Parthenon to be enlarged. Carpenter has stated that the Parthenon, as built, is wider than the original plan but that the columns were taken from the Kimonian Parthenon and reused; their spacing was altered from 14 ft 4 in. (4.4 m) to 14 ft 0 in. (4.3 m), as determined by the space available for the platform. Many archaeologists consider that the evidence is insufficient to support Carpenter's supposition, but there is no adequate evidence either to credit the view that the column spacing was based on the Golden Section (Ref. 1.3, Section 10.8) or on any other proportional rule.

The masonry of the Parthenon, like that of most Greek architecture, was laid without mortar (Fig. 3.13). The joints were carefully fitted; probably by grinding the contact surfaces on sand. The individual drums of the columns were joined with hardwood or iron or bronze dowels which resisted the horizontal shear (Fig. 3.14a and b).

Both in the Parthenon and the Theban Treasury at Delphi blocks of stone were attached to one another horizontally with iron clamps, a technique that may also have been applied in other Greek temples. It was later extensively adopted in ancient Rome (Section 3.9) and the Renaissance (Section 7.2).

The Greeks were more economical in their handling of materials than the Egyptians had been and the individual blocks of stone they used tended to be smaller. In particular, it was common practice to build the architraves (horizontal beams) from two parallel blocks of stone (Fig. 3.15).

It seems likely that Greek temples were once built entirely of timber and that the actual roof in many remained a timber structure. Hence the Parthenon and other Greek temples today generally consist of no more than a colonnade, the roof timbers having perished. Their sizes and spans are not known with certainty, but many classicists have produced reconstructions (e.g., Ref. 3.23, pp. 172 and 184). The sizes can sometimes be guessed from the notches that remain in the stonework,

3.13

The ruins of the Parthenon in Athens. The masonry was carefully finished by rubbing with sand and the drums of the columns were joined by wooden dowels, later by iron dowels. No mortar was used.

although they give an upper limit; the timber sections may have been smaller than the notches. Some temples, however, are roofed wholly or partly in stone (Fig. 3.16).

Just as the timber has perished, so has the iron been destroyed by rust or plunder. Dinsmoor (Ref. 3.10) has stated that the Greeks used iron as a reinforcement (see also Section 3.9), but no definite evidence supports this claim. Iron was employed (Refs. 3.24 and 3.25) in the architraves of the Propylaea, the gateway to the Acropolis of Athens, but it was placed on the top of the stone beams where it was compressed (stressed in compression). It was probably used *under* the architraves of the Temple of Zeus in Agrigento in Sicily, which was never completed; it may, however, have served primarily as a construction aid or permanent formwork.

The Parthenon remained intact until the fifth century A.D. when the statue of Athena was removed and the building dedicated as a church to S. Sophia. In 1458 the Turks turned the building into a mosque without material damage, except that a minaret was added. In this form the building has been described by several travelers in the seventeenth century. During the siege of Athens by the Venetians the Turks used the building as a powder magazine which blew up in the bombardment.

3.14a

Fallen column from the temple of the Olympian Zeus in Athens. Construction was begun in 174 B.C., but finished by the Roman Emperor Hadrian in A.D. 117. The drums were joined with wooden dowels to resist the horizontal shear. The dowels have vanished but the square holes are still visible.

After the Venetians captured the city General Morosini damaged the ruins further by attempting to remove the horses of Athena from the western pediment. The best of the remaining sculptures were taken to England in the early nineteenth century by Lord Elgin, then British Ambassador to Turkey, and are now in the British Museum.

In spite of the structural limitations, the Doric, Ionic, and Corinthian orders developed by the Greeks between the sixth and fourth centuries B.C. have influenced architectural design, except for the medieval interlude, up to and into the twentieth century, and, by way of Europe, the architecture of other continents (Section 8.8).

Greek architecture came in for severe criticism from Neo-Gothic theorists for its inefficient use of stone. Heyman (Ref. 3.24, p. 3) quotes Ruskin as saying in 1853 that

the Greek system, considered merely as a piece of construction is weak and barbarous, compared with the other two (Roman and Gothic). For instance, in the case of a large window or door, if you have at your disposal a single large and long stone you may indeed roof it in the Greek manner with comparative security; but it is always expensive to raise to their place stones of this large size, and in many places nearly impossible to obtain them at all.

3.14*b*

Drum from the same column with the remains of iron dowels. The rusting of the iron has cracked the stone.

This is a valid criticism, and it was the Roman development of the arch (already known to the Egyptians and Greeks) that dominated the design of all great interior spans until the nineteenth century. It is only fair to point out, however, that by 1853 iron was posing a new challenge to the traditional masonry structure because it enabled designers to build long spans by using merely the horizontal and vertical members favored by the Greeks; Ruskin himself was open to criticism for failing to appreciate the significance of the new engineered structures.

In actual fact the Greek stone beams have a considerable reserve of strength above that of a simply supported beam. Because of their weight, the frictional forces are generally sufficient to prevent horizontal movement so that the beam acts like a flat arch.

The problem has been examined by Heyman (Ref. 3.24) in relation to the Temple of Aphaia on the island of Aegina (Fig. 3.17). This was described by Banister Fletcher (Ref. 3.20) and detailed photographs have been published by Mussche (Ref. 3.2). The temple is built of a soft yellow limestone, originally coated with a thin stucco and partly painted. The architrave measures 850 mm by 850 mm (2 ft 9 in. by

3.15

Architrave of the Theseion in Athens built up from two parallel blocks of stone.

3.16

Roof of the Theseion on the Old Agora (market place) in Athens. It was probably a temple of Haphaestos built about 465 B.C.

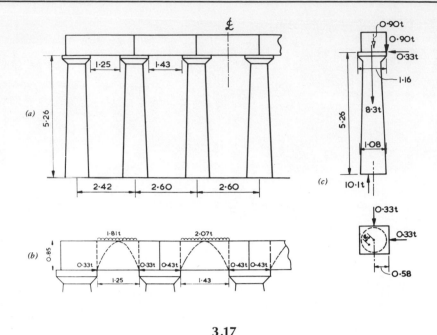

3.17

Temple of Aphaia on the island of Aegina.

(a)   Principal dimensions in meters.
(b)   Possible forces in the existing architrave, in the old metric unit of tonnes-force (1 t = 1000 kgf = 2205 lbf = 9.81 kN).
(c)   Forces in the corner column in tonnes-force. The eccentricity e was determined by Heyman as 0.16 m (6 in.).

[From Ref. 3.24. Copyright © 1972 by the Society of Architectural Historians (all rights reserved)]

2 ft 9 in.) and the clear span is 1.43 m (4 ft 8 in.), except for the shorter end span of 1.25 m (4 ft 1 in.). Taking the weight of the limestone as 200 kg/m³ (125 lb/cu ft), the bending moment over the larger span is 3.6 kNm (2560 lb-ft), which produces a tensile stress of 35.40 kPa (5 psi) in the limestone (see Section 8.3). This is probably no more than one-thirtieth of its tensile strength.

If cracking occurs because of an earthquake or temperature movement of the columns, the beams still have a substantial margin of safety. Heyman considered that the cracked beams act like a flat arch (see Section 7.6) and that the bending moment would be resisted by horizontal thrusts which set up a reverse moment. Heyman calculated the critical force in the end span as 3.2 kN (719 lb) which could be transferred to the column by friction between the architrave and column. The end column now has a horizontal load of 3.2 kN (719 lb) as well as the vertical load, and Heyman calculated (Fig. 3.17) the resulting eccentricity as 0.16 m (6 in.) on a column with a base diameter of 1.8 m (3 ft 7 in.). The structure therefore appears safe even if the architrave is cracked.

Structure in the Ancient World

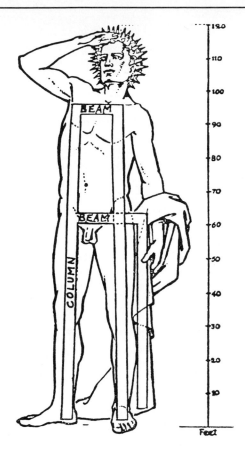

3.18

Reconstruction of the Kolossos of Rhodes by Herbert Maryon in 1956 (Ref. B 98).

The most daring Greek structure was the Kolossos which stood at the entrance of the harbor of the island of Rhodes and was classed as one of the Seven Wonders of the World (Section 3.1). Hamilton (Ref. 3.25) quoted an account by Maryon of the British Museum, based mainly on the writings of Philon of Byzantium (ca. 200 B.C.), Strabo (ca. 53 B.C. to A.D. 21), and Pliny (A.D. 23 to 79), supplemented by his own conjectural restoration (Fig. 3.18):

A plinth of white marble 20 to 25 ft (6 to 7 m) high was built. On this the statue stood, according to Philon, 70 cubits high. Taking a cubit as 20.7 in (526 mm), Maryon gives the height as about 120 ft (36 m). Within the legs and the drapery that hung from the left forearm three great cylindrical drums were raised, probably without mortar as was the Greek custom, but strengthened in some way by iron bars. In his reconstruction Maryon supposes the iron bars to have been 4½ sq in. (2900 sq mm) in cross section at the ankle level of the figure, tapering both above and below that level, and let into chases in the columns which probably filled the metal casing up to ankle level. At that level and above they would be about 5 ft (1.5 m) in diameter. At 60 ft (18 m) above the plinth Maryon shows the columns at pelvis level tied to one another by beams, then two of them carried up within the body to shoulder level at about 90 ft (27 m) and again tied by a beam.

To bring material to the working level, according to Philon, a ramp wound its way round a vast mound of earth that was piled about the statue as it rose. The iron used in the construction weighed 7½ tons and the bronze 12½ tons. Comparing that weight and surface with those of a life-size statue, Maryon computed the thickness of bronze to have been 1/16 in. (1.6 mm). It must in that case have been of sheet beaten to shape and not cast. If the sheets were about 3 ft (1 m) high they would need profile bars, say iron flats on edge running around the statue horizontally at vertical intervals of 3 ft. These would be held in place by iron stays of which the inner ends were built into one of the masonry columns. Above shoulder height the profile bars for neck and head would be held in place by attachment to some kind of iron framework.

About 224 B.C., 56 years after its completion, the Kolossos was thrown down by an earthquake. The plinth and the legs up to the knees remained standing; above the knees the figure bent over until the head and shoulders came to rest upon the ground, exposing through rents in the covering fallen masonry and bent ironwork, and so the wreck remained until A.D. 653 (i.e., for 877 years) when the Saracens arrived, broke it up, and sold the metal to a scrap dealer.

3.3 ANCIENT ROME

Rome was founded in the eighth century B.C. The Roman conquest of Italy began in the fourth, and the victory over Carthage in the three Punic Wars (264 to 146 B.C.) made Rome a great power. It then conquered all the Mediterranean countries and all of western Europe, except Ireland and northern Scotland.

The friction between the generals in command of the conquering legions led to a breakdown in the republican government and in 27 B.C. Octavian declared himself emperor with the title Augustus. The government of the Empire gradually deteriorated during the third and fourth centuries A.D., and in A.D. 395 the Empire was divided into eastern and western sectors. The Western Empire came to an end in A.D. 476, but the Eastern (Byzantine) Empire continued until the Turks conquered Constantinople in 1453.

Rome was notably more tolerant than the Greek city states which limited democratic rights and professional status to a minority of free citizens by excluding slaves, serfs, and foreigners. In the Roman Empire it was possible for a man of any race, even a former slave, to achieve high professional standing. The Romans, however, even more than the Greeks, depended on manpower provided by slavery.* After the first century B.C. they left the determination of who should rule to the force of arms.

The Roman conquests created three centuries of peace during which the armies were engaged in only minor skirmishes at the frontiers. This is the longest period without a major war that civilized Europe has known, and the *pax romana* made it possible to direct energies to projects never before attempted.

The Roman contribution to architecture was distinguished by its magnitude and technical sophistication. Roman remains and buildings are still in use from Spain to

* Until the third century A.D., which includes the period during which the architecture discussed in this chapter was built; thereafter the Romans gradually developed the *colonus*, the forerunner of feudal serfdom.

Asia Minor and from North Africa to Britain (Refs. 3.19, 3.20, 3.23, 3.27, and 3.28). The quality of the concrete construction, water supply, and heating systems remained unequaled until the nineteenth century.

The most impressive Roman architecture dates from the last years of the Republic to the reign of the Emperor Caracalla, that is, the first century B.C. and the first, second, and early part of the third centuries A.D. At its best Roman architecture was simple, functional, and well proportioned. It was less successful when it imitated Greek architecture.

Roman architects and engineers were not markedly inventive, but they appreciated a good idea when they saw one and experimented with it until it was technically perfect. The Romans did not invent concrete or the arch or water supply or sewage disposal, but they were the first to build great interior spans, the first to build public baths, and the first to build sewage disposal and water supply systems for the entire population. In the process some magnificent structures were created.

The extent of the Roman achievement is illustrated by the decline in the design of buildings following the withdrawal of the Roman administration from Africa, Spain, France, and Britain (Refs. 3.4 and 3.6). This was a gradual process, and it is still a matter of surprise that Roman methods of construction, for example, the hypocaust (Section 4.6) so ideally suited to heating buildings in Britain, were completely forgotten for more than a thousand years.

3.4 ROMAN TIMBER STRUCTURES

No Roman timber structures survive, but we have descriptions of some; furthermore, holes left by timber posts enable us to guess at the construction of the buildings.

In Book 4, Chapter 7, of *De Bello Gallico* (the commentary on his wars in France and Germany) Julius Caesar described the construction of a timber bridge across the Rhine near Coblenz, about 400 m (¼ mile) long, in a mere ten days. He lowered timber piles, 450 mm (18 in.) square, into the river from rafts and drove them into the river bed with a *machinatio*, presumably a pile driver that utilized a dropping weight not unlike those still in use. These piles were placed in pairs at a distance of 12 m (40 ft) and between each pair a transom was fixed. The bridge deck spanning 12 m (40 ft) was laid across these transoms. There are no contemporary illustrations, but in the sixteenth century Palladio (Ref. 3.26, p. 62) made the drawing shown in Fig. 3.19.

A contemporary illustration of the bridge built by Apollodorus over the Danube in Dacia just below the Iron Gates (the modern border of Yugoslavia and Romania), appears on Trajan's column still standing in Rome. It shows the stone foundations and a timber superstructure with crossed diagonals (Ref. 3.17, p. 144).

We can only guess at the nature of Roman timber buildings. Richmond has discussed the subject in the light of British excavations (Ref. 3.5) and Rickman (Ref. 3.27, p. 241) illustrated the reconstruction of a timber *horrea* (storehouse), measuring 48 by 30 m (157 by 97 ft), excavated in Roman Germany.

In addition to the archaeological evidence we have also the description of timber construction by Vitruvius. The timber flooring is to consist of winter oak boards to avoid warping. It is then to be covered with fern, if there is any, or with straw to protect the wood from being damaged by lime. The top layer is a mixture of small

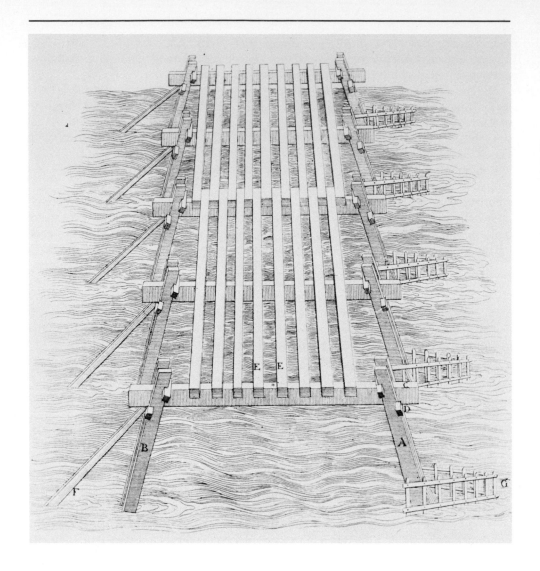

3.19

Palladio's sixteenth-century illustration of the bridge built by Julius Caesar over the Rhine in 55 B.C. (Ref. 3.26, p. 62).

stones, lime, and pounded tile, which is rubbed down after setting. (Ref. 2.3, pp. 202–203).

When vaulting is required, the procedure should be as follows. Set up horizontal furring strips at intervals of not more than two feet apart, using preferably cypress, as fir is soon spoiled by decay and age. Arrange these strips so as to form a curve, and make them fast to the joists of the floor above or to the roof by nailing with many iron nails to ties fixed at intervals. . . . Having arranged the furring strips, take cord made of Spanish broom, and tie the Greek reeds, previously pounded flat, to them in the required contour. . . . Having thus set the vaulting in their places and interwoven them, apply the rendering coat to their lower surface; then lay on the sand mortar, and afterwards polish it off with powdered marble (Ref. 2.3, Book 7, Chapter 3, pp. 205–206).

Faventinus, writing almost three centuries later, offered an almost identical prescription (Ref. 3.29, Sections 18–19, p. 67).

This type of construction could not, of course, have survived to the present time, but because it is the one recommended by both authors it was presumably the normal method of construction. It is quite likely that the greater part of the construction in Greece and Egypt was also of timber.

3.5 ROMAN CONCRETE

We are on much firmer ground in discussing Roman concrete because the remains are extensive.

The Romans did not invent concrete, for sands of volcanic origin with cementing properties are found in many parts of the Mediterranean. When mixed with lime, they form a mortar with properties not unlike those of modern cement mortar. One notable example is the Greek island of Thera (better known as Santorin, a name it received during the Fourth Crusade) on which the local volcanic soil is still a dominant building material (Ref. 3.7). Lime concrete was used by the Etruscans, from whom the Romans may have acquired the technique. Concrete appeared in Rome probably during the second century B.C., that is, after the Second Punic War. It is mentioned by Cato (232–147 B.C.) and described by Vitruvius, but not recommended for extensive use. In the days of Augustus (27 B.C.–14 A.D.) builders apparently had little confidence in concrete because of the extensive use of concealed voussoirs as back-up reinforcement. Excavations and buildings still standing show that by the first century A.D. it had become an important material and continued to be until the fourth century, that is, about the time of Constantine the First, when it declined in favor.

In Imperial Rome concrete construction reached a degree of sophistication it did not regain until the nineteenth century (Refs. 2.3 and 3.29 to 3.34). This was probably due to a combination of circumstances: the great skill of the Romans in construction and the ready availability near Rome of volcanic sand with cementing properties which was used for *opus signinum*, a mortar containing broken potsherds or bricks, and *opus caementitium,* a mortar cast around large pieces of natural stone or rubble from the demolition of older buildings and used for the enormous structures that have proved so durable.

For particularly important work *pulvis puteolanus* was imported from the Roman port of Puteoli (the modern Pozzuoli, near Naples). This material has become famous under the name pozzolana, a term still used for certain cements (although with different properties). Vitruvius, who wrote in the first century B.C., described it as follows:

There is also a kind of powder which from natural causes produces astonishing results. It is found in the neighbourhood of Baiae and in the country belonging to the towns around Mount Vesuvius. This substance, when mixed with lime and rubble, not only lends strength to buildings of other kinds, but even when piers are constructed in the sea, they set hard under water. The reason for this seems to be that the soil on the slopes of the mountains in these neighbourhoods is hot and full of hot springs. This would not be so unless the mountains had beneath them huge fires of burning sulphur or alum or asphalt. So the fire and the heat of the flames, coming up hot from far within through the fissures, make the soil light, and the tufa found is spongy and free from moisture. Hence the three substances, all formed on a similar principle by the force of fire, are mixed together, the water taken in makes them cohere, and the moisture quickly hardens them so that they can set into a mass which neither the waves nor the force of water can dissolve. (Ref. 2.3, Book 2, Chapter 6, pp. 46–47).

Vitruvius' explanation of the behavior of cementing substances in terms of the four elements of Aristotle (Section 2.2) does not agree with the findings of modern chemistry. It does, however, draw attention to an important distinction between lime mortar, used throughout the Middle Ages, the Renaissance, and into the nineteenth century, and cement mortar. Lime mortar is water soluble and thus washed out by water. A "hydraulic mortar," made with pozzolana and lime or with modern portland cement, is waterproof and can be used for hydraulic works.

It seems likely that the "pitsand" (fossiciae) originating from extinct volcanoes near Rome also had hydraulic properties. Vitruvius described it in Book 2, Chapter 6, Section 3: "If it lies unused too long after being taken out, it is disintegrated by exposure to sun, moon and hoar frost, and becomes earthy"; this suggests that, unlike "river sand," it contained alumina.

Roman concrete was not, however, dependent only on volcanic "sand." I had the opportunity shortly after World War II to test some concrete from Roman ruins in Libya. Tests on 3-in. (76-mm) cubes indicated a strength of 2.8 ksi (19.3 MPa). Chemical analysis showed that the material consisted only of silica sand, hydrated lime, and broken brick; there had been some fusion between the last two. Presumably a good deal of *opus signinum* was of that kind. Vitruvius, Faventius, and Palladius (Ref. 2.3 and 3.29) all prescribed the addition of a third part of crushed earthenware to mortar when "river sand" was used, which suggests that it was intended to act as a substitute for pozzolana when the chemically inert river sand (i.e., silica sand) was employed.

The relatively high strength of Roman concrete may be partly due to age, but it probably owes more to a low water-cement ratio. Much of the poured Roman concrete contains tufa, selce, and other porous rocks of volcanic origin which absorb some water. Moreover, Pliny adjured the mason to be careful with the use of water.

The Romans were well acquainted with the effects of low and high temperatures on concrete. Frontinus stated (Ref. 3.30, p. 352) that repairs on aqueducts should be carried out only between the first of August and the first of November and also suspended during very hot weather.

Somewhat confusingly for the modern reader, the Latin word for concrete aggregate is *caementum. Opus caementitium*, or concrete, thus contained pieces of aggregate. Roman aggregate varied greatly in its composition and size; for unimportant work broken stone or concrete was obtained from the demolition of old buildings but for important buildings the aggregate was carefully selected. The heavier materials were used at the bottom and especially lightweight aggregates such as pumice higher up. Pieces were generally much larger than in modern concrete and placed in position in regular layers before the mortar was poured in between.

3.20a

The Pantheon in Rome built approximately A.D. 123 and still in use as a church. The span of 43 m (143 ft) was unsurpassed until the fifteenth century (42 m or 138 ft in the Duomo of Florence) and by no other before the nineteenth century. The construction is massive; the walls are 7 m (23 ft) thick between relieving arches. Eighteenth-century etching by Piranesi.

The Pantheon (Fig. 3.20) has a span of 43 m (143 ft), the longest to have been constructed before the nineteenth century, which made the relative weight of the caementa very important (Ref. 3.32). The lowest part was built with aggregate of broken brick, which was changed to alternate layers of brick and tufa (a porous volcanic rock). The upper part of the dome was built with alternate layers of tufa and pumice, the latter imported from Mount Vesuvius to reduce the weight of concrete.

The concrete is exposed on the inside of the Pantheon; however, the outside, like most Roman concrete structures, is faced with brick, placed both as permanent formwork and as a veneer. The Romans rarely used exposed concrete on the outside of buildings.

The marks of the timber forms are clearly visible on some buildings; for example, the octagonal dome of the *Domus Aurea*, the Golden House of Nero, still to be seen

---

PANTHEON
SEZIONE =

Structure in the Ancient World

in a park near the Colosseum in Rome (Ref. 3.33). This is one of the few surviving large Roman concrete domes, although it is much smaller than the Pantheon (14 m versus 44 m) and was built half a century earlier.

3.6 FORMWORK FOR CONCRETE

Roman masonry served two purposes: the conventional one of providing a structure of brick or stone in its own right and a new use as permanent formwork to concrete construction. During the early years of the Empire concrete became an important structural material.

The Romans were no more concerned with structural sizes than the civilizations before them (Section 3.9); they did, however, pay careful attention to ease of construction. In this respect their thinking was not unlike that customary today. We also use massive concrete construction for large structures because it reduces the labor content and thus the cost.

The Romans had less reason to economize on labor than preceding civilizations. While the security of the Greek city states was constantly at risk, the *pax romana* of the first and the second centuries A.D. allowed the diversion of large resources to building. The Romans, however, in spite of the extensive use of forced labor, may well have been motivated by a shortage of manpower, caused by the size and the speed of the building programs undertaken during those two centuries (Fig. 3.21); Roman buildings were completed much faster than the Gothic cathedrals and generally much faster than the buildings of the Renaissance.

The strength of Roman concrete structures presented no problems *after* the concrete had hardened. The structures were massive; for example, the walls of the Pantheon were 7 m (23 ft) thick (Fig. 3.20), but the pressure of the liquid mortar had to be resisted *until* it hardened. Because of the Roman practice of laying the caementa in regular layers so that they supported some of their own weight during construction, the liquid pressure of the concrete would not have been so great as that of an equal volume of modern concrete containing small aggregate as part of the mix; however, the pressure of the mortar alone must have been considerable.

To resist this pressure the Romans used masonry as permanent formwork and included arches to increase its strength wherever there was a change of cross section (Fig. 3.22a). Horizontal brick courses were sometimes placed in the concrete to act as ties while the concrete was still wet (Fig. 3.22b).

A certain conservatism may also have contributed to the use of brick and stone as permanent formwork. People were used to seeing brick and stone walls and continued to see them even after concrete had become the dominant building material in the first century A.D. The Roman caution was probably justified. Most concrete surfaces of the nineteenth and early twentieth centuries leave a good deal to be

3.20*b*

Hypothetical cross section of the Pantheon.

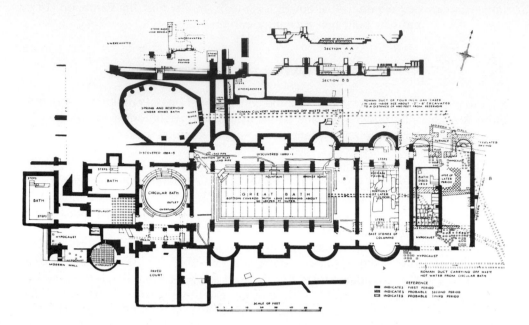

3.21

Plan of the Roman Baths at Bath, England by A. J. Taylor. Recent excavations show the elaborate nature. (Ref. 3.4). This great structure houses a lead-lined warm bath about 21 m (70 ft) long, three smaller baths for higher temperatures, rooms with underfloor (hypocaust) heating, and an efficient system of waste disposal, yet it was a provincial bathing establishment in the most remote Roman diocese.

3.22a

Concrete from the Baths of Trajan in Rome, exposed by the collapse of the permanent formwork. Note the impression of the brick relieving arches; one is still partly in position.

3.22*b*

Horizontal brick courses which acted as stiffeners during construction of the Baths of Trajan.

3.23

Ruins of the Basilica of Maxentius in Rome, built in A.D. 312 as a law court. The vaults span 25 m (83 ft) and the crowns are 40 m (120 ft) above the ground. The walls supporting the vaults are massive, but arched communicating openings are formed in them. Note the coffers on the inside of the vault.

desired, and Rome lacked the technology that has enabled us to solve the problem in the last thirty years (Ref. 1.3, Section 9.5).

The Romans used lightweight aggregate to reduce the weight of the higher parts of the Pantheon (Section 3.5) and other buildings. In addition, there are coffers in the roofs of the Pantheon (Fig. 3.20) and the Basilica of Maxentius (Fig. 3.23) that cut down the amount of concrete to some extent. Empty wine and oil jars of clay have been found cast in the concrete. It was at one time thought that this might be just careless construction, but hollow clay tubes apparently made specifically for this purpose have also been found in Nero's Golden House and in houses along the Via Appia (Ref. 3.33). Rectangular hollow clay tiles have been found in Bath (Fig. 3.24).

3.24

A fragment of roofing from the Great Bath at Aquae Sulis (the Roman city of Bath, England), shows box-tile vaulting and an external finish of ridge tiles set in cement mortar.

3.7 ROMAN MASONRY AND BRICKWORK

Most of the best known masonry structures of ancient Greece, Mexico, and Peru were built without mortar. The carefully fitted joints must have required a great deal of labor. The Inca masonry surviving in Cuzco, Peru, has mortarless joints so tight that it is impossible to push the blade of a knife into them (Fig. 3.3b and c). This could have been achieved only by grinding every joint with sand until the stones fitted perfectly. Although the craftsmanship of masonry without mortar is wholly admirable, the expenditure of labor on stones and bricks laid in mortar is obviously much lower. Mortar joints therefore are common in Roman masonry, but they coexisted with dry construction (Figs. 3.7a and 3.28).

The Romans occasionally employed mud as a mortar, particularly for laying sundried bricks. Pure lime mortar was also used; however, the addition to lime of crushed tiles or a pozzolanic material such as pit sand of volcanic origin was normal practice (see also Section 8.7). This was one reason for the durability of Roman construction; another was the care paid to selecting durable materials (Section 4.1).

Structure in the Ancient World

3.25

This brick vault over a cistern on the slope of the Acropolis in Athens, dates from the late Roman period.

3.26

Opus testaceum, concrete with permanent formwork of bricks cut diagonally.

Structure in the Ancient World

3.27

Opus quadratum, squared masonry blocks, form part of the Temple of Jerusalem built by King Herod about 34 B.C.

Pliny recommended a ratio of one part of lime to four parts of "pit sand" (Ref. 3.30, p. 316), whereas Vitruvius recommended 1:3.

Roman bricks were mostly longer and wider and sometimes thinner than modern bricks (Fig. 3.25). Marion Blake (Ref. 3.30) reported brick sizes up to 500 by 300 mm (20 by 12 in.). Thicknesses ranged as low as 12 mm (½ in.); thin bricks were easier to bake without warping or cracking.

Bricks were bonded, as in current practice, by headers and stretchers or bricks of double width were stretched across two courses. Bricks used for concrete formwork were frequently cut diagonally (Fig. 3.26). This was called *opus testaceum* and the earliest recorded use was in Nero's Golden House.

Roman stone walls were laid as *opus quadratum* (Fig. 3.27) with conventional horizontal and vertical joints, as *opus reticulatum*, carefully squared and laid with joints at 45° to the horizontal in a diamond-shaped bond, or as *opus incertum*, stones only roughly squared and laid in the same way. Vitruvius (Ref. 2.3, Book 2, Chapter 8, p. 51) preferred the last:

There are two styles of wall: *opus reticulatum*, now used by everybody, and the ancient style called *opus incertum*. Of these the reticulatum looks the better, but its construction makes it likely to crack, because its beds and builds spread out in every direction. On the other hand, the opus incertum, the rubble lying in courses and imbricated makes a wall which, though not beautiful, is stronger than the reticulatum.

Notwithstanding Vitruvius' criticism, a great deal of opus reticulatum still stands, some as permanent formwork for concrete. It is a distinctive bond characteristic of ancient Roman masonry, whereas opus quadratum and opus incertum have a more "modern" appearance.

3.8 TALL BUILDINGS AND FIRE PRECAUTIONS

Estimates of the population of ancient Rome at its peak in the late third and early fourth centuries vary from as low as one-quarter million (Ref. 3.35, p. 201) to as high as two and one-half million (Ref. 4.6, article "Rome"), which would make it larger than any other city before 1850; possibly one million is a reasonable figure but it is still higher than the population of London in 1800. As Rome grew, more and more of its people were accommodated in *insulae*, blocks of flats that occupied whole blocks or "islands." Camp (Ref. 3.35, p. 169) stated that in the year A.D. 300 a census of Rome listed 1797 domus (houses) and 46,602 insulae (i.e., 26 times as many). Only the lower floors of insulae have survived (e.g., in Ostia Antica), but we know that after the great fire of A.D. 64 the Emperor Nero issued a decree limiting the height of buildings to 21.5 m (70 ft), which suggests that some must have been higher. Evidently the people on the upper floors had to climb a lot of stairs.

Other buildings were even taller; for example, the Colosseum, the great Flavian amphitheatre of Rome, which is still standing, has a height of 48 m (158 ft) (Fig. 3.28). Before the nineteenth century the limit was set by vertical transportation.

The high density of Rome seems to have created a serious fire hazard; but fire fighting became well organized during the first two and one-half centuries of the Empire. We first hear of groups of slaves employed as fire fighters in the first century

3.28

The Flavian Amphitheater in Rome, commonly known as the Colosseum, built between A.D. 70 and 82. It is 48 m (158 ft) high and the tallest building surviving from ancient Rome. Blocks of travertine, laid without mortar and held together by clamps, covered the facade. Although it was used for centuries as a quarry for building stone parts of it are still in excellent condition. See also Fig. 4.6.

B.C. After approximately a quarter of Rome was destroyed in the fire of A.D. 6 the Emperor Augustus created the corps of *Vigiles*, a professional fire-fighting service whose officers held army rank and whose chief, the *Prefectus Vigilum*, ranked immediately after the Prefect of the Imperial Guard.

The fire fighters were trained in specialized tasks (Ref. 3.11). The *Uncinarius*, or hook man, pulled away burning roofs before they collapsed into the interior of a building. The *Aquarius* was responsible for the water supply and for transporting water in buckets to the site of the fire. The *Siphonarius* actually directed the water onto the fire. Recent excavations at Silchester, England, and Civitavecchia, Italy, have produced bronze parts that indicate the existence of elaborate pumps containing cylinders, pistons, and valves which may have been capable of ejecting water with considerable force (Ref. 3.11, p. 13). Fire brigades apparently were established throughout the Empire as part of the normal administration. The existence of a good water supply (see Section 4.3) was an important aid to fire fighting, but once the fire got hold it was stopped by the wholesale demolition of buildings to create a firebreak (Ref. 3.30, p. 42).

Among the several accounts of the fire of A.D. 64 in the reign of the Emperor Nero that of Tacitus is the best known. Whatever Nero may have done during the fire, he acted decisively to minimize future danger. Suetonius described some of the fire precautions. Each new insula was required to have its own separate wall; party walls were forbidden. Porticoes were required to have accessible flat roofs for fire fighting. Each householder had to have in the open his own fire-fighting apparatus. We have already mentioned the limitation on the height of buildings. The use of timber in building was restricted. Nero also legislated that *saxum gabinum* and *saxum albanum* (stones now called *sperone* and *peperino*) be used whenever possible because of their superior resistance to fire, but because of the cost of overland transport from the Alban hills this law proved difficult to enforce.

Nero succeeded where Wren failed after the Great Fire of London in 1666 (Section 7.9). He cut straight wide roads through the burned-out parts of the city, some of which are still arterial roads in present-day Rome; for example, the Via del Corso. These roads were intended to act as firebreaks and to provide access for fire fighting.

Nero came in for considerable criticism when he built himself a great palace, the *Domus Aurea* (Golden House), on a burned-out area between the Palatine and the Oppian Hill, using rubble from the destroyed buildings for concrete aggregate.

3.9 ROMAN ARCHES, VAULTS, AND DOMES

We now return to the subject we considered at the beginning of this chapter, namely, structural design. The great contribution of the Romans was the development of the arch, the vault, and the dome. They did not invent these forms (Section 3.1) but they perfected them to an extent that was unequaled until the fifteenth century and unsurpassed until the midnineteenth.

The distinction between the true arch and the corbeled arch is not so sharp as sometimes imagined. During the nineteenth century the concept developed that the corbeled arch consists of a series of miniature cantilevers balanced on top of one another and that it does not produce a horizontal thrust. This is true only if the corbels are able to slide on top of one another.

Structure in the Ancient World

In Roman concrete the caementa were often arranged in horizontal layers (Section 3.5 and Fig. 3.20b), and this at first suggests a corbeled structure. However, the cement mortar in domes, such as the Pantheon, was very strong and the sliding action was firmly resisted. Thus it behaves like any normal concrete structure and is a true dome.

Even if the mortar is less strong, as in the Ctesiphon vault (Section 5.3), true arch action occurs if the weight of the material produces frictional forces between the layers of blocks or bricks strong enough to stop movement of the horizontal layers in relation to one another. It follows, of course, that the structure produces horizontal reactions at ground level.

If the structure is relatively light, the weight may be insufficient to produce friction in the horizontal joints, and the corbels then act as individual cantilevers whose span is limited by the tensile strength of the material.

The true arch is therefore a distinct structural advance, particularly if we wish to reduce the weight of the structure.

The Roman masonry arch was normally a true, not a corbeled, arch. Consequently it needed formwork to support it during construction which a corbeled arch does not require. The true arch became a self-supporting structure when the last stone was inserted at the top. This was called the keystone and its placement was frequently accompanied by a religious ceremony. The Romans often decorated the keystone, which seems appropriate, and sometimes used a stronger material which does not, for the keystone actually carries the smallest load (Section 7.3).

The Romans frequently used concrete for arches after the first century B.C. because of the saving in cost, but extra formwork was needed to carry the weight of the wet mortar until it hardened (Section 3.6). It was therefore rarely used for bridges because of the difficulty of erecting the heavy formwork across water.

The Romans built numerous arched bridges; some in Italy, France, and Spain are still in daily use. They also built aqueducts (Section 4.3) that involved the construction of water channels as high as 54 m (180 ft) above the bottom of the valley and required three tiers of arches. Some of these structures are still standing and in spite of their great height are in harmony with the landscape. They built many kilometers of vaulting in their theaters and big-span vaults in their baths and basilicas (Fig. 3.23). Perhaps the most interesting structures are the domes, among which the Pantheon is the most important (Fig. 3.20), yet we know nothing of the design of these structures. No Roman author (Section 2.1) mentioned the subject.

Roman arches, vaults, and domes were almost invariably semicircular. The interior space of the Pantheon was also designed as a sphere (Fig. 3.29). The question why the Romans used the semicircle to this extent is perhaps best answered by "Why not?" The circle dominated Euclidean geometry; it was described by Greek philosophers as the perfect curve and was the easiest to set out. The Romans would naturally have assumed that it was also the best for strength, although there may have been evidence in Mesopotamia to the contrary even in their time (Section 5.3).

The problem is somewhat different for the arch and the vault on the one hand and the dome on the other. Because of the double curvature, the dome has more stability and is therefore suitable for larger spans.

As we shall see later (Section 7.6), arches of stone or brick fail through the opening of four joints. The Romans may have been aware of this fact from experience, and

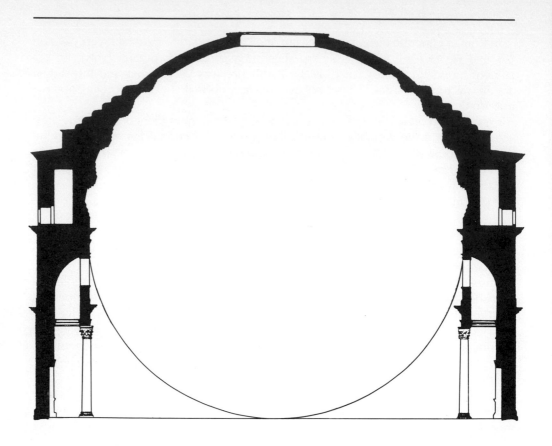

3.29

Circle inscribed in the cross section of the Pantheon (Ref. 3.8, Fig. 11).

metal clamps, usually of iron, sometimes of bronze, are found in a number of masonry arches, particularly in dry-jointed masonry. In some cases every stone is fixed to every other (see also Fig. 8.3) by two metal clamps; for example, in the Pons Cestius over the Tiber in Rome, which is now called the Ponte Cestio and still in use (Ref. 3.36, p. 12). Sometimes the effect of the metal reinforcement is almost like that of reinforcement in modern concrete.

The horizontal reactions, or thrusts, of adjacent arches balance, and it is necessary only to provide a more substantial support at the end of the arcade or line of arches. When a wall supported by arches collapses, the arches often remain standing as a self-supporting structure. It may be that this observation led the Romans to erect arcades supported on slim columns. The first of these is found in the palace erected by the Emperor Diocletian in the third century A.D. at Spalatum in the diocese of Italy (Split in present-day Yugoslavia). This type of construction, later adopted by Christian architects, dominated the design of Romanesque and Gothic churches for centuries.

The problem of absorbing the horizontal reaction of a dome is different. The vertical "arching" stresses are compressive throughout, but the horizontal stresses

Structure in the Ancient World

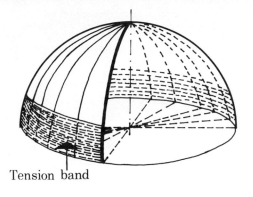

Tension band

3.30

Vertical "arching" stresses and horizontal "hoop" stresses in a thin hemispherical dome. The arching stresses are compressive throughout. The hoop stresses are compressive in the upper portion of the dome and tensile in the lower. The change occurs at a circle which forms an angle of 52°24′ with the crown and 37°36′ with the horizontal (Ref. 3.9, p. 324).

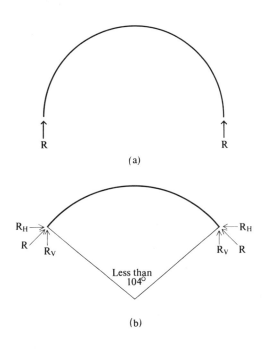

(a)

(b)

3.31

Reactions of (a) the hemispherical dome and (b) the shallow dome.

3.32

Cracks in the concrete masonry of a semidome in the Baths of Trajan at Rome. The iron bar is a modern reinforcing bar.

are tensile in the lower portion if the dome is a complete hemisphere (Fig. 3.30). In the Pantheon (Figs. 3.20 and 3.29) the thickness of the concrete is so great in the lower portion of the dome that the tensile stresses are low.

Semicircular arches and vaults produce horizontal reactions, or thrusts, which must be resisted externally. In a hemispherical dome, however, these horizontal reactions are absorbed by the hoop tension. The reactions of the hemispherical dome are therefore purely vertical, and even if the dome is high above ground they can be transmitted easily downward (Fig. 3.31a).

Hoop tension can be avoided and the dome made much thinner if it is made so shallow that its lower part in which the hoop tension occurs is eliminated.

This produces an inclined reaction, or thrust, which has both a vertical component R_V and a horizontal component R_H (Fig. 3.31b). The horizontal component must be absorbed by buttresses. Thus we avoid one problem, the hoop tension, and create another, the need for buttressing.

The hemispherical dome is characteristic of western Roman and Renaissance architecture. The shallow dome is characteristic of eastern Roman (Byzantine) and Muslim architecture.

It should perhaps be pointed out at this stage that in modern reinforced concrete construction neither the hemispherical nor the shallow dome presents any serious structural problem; the modern preference for the shallow dome is for functional reasons.

Because of the thickness of Roman domes, the cracks observed in some of them (Fig. 3.32) are probably due to temperature movement, or to movement caused by the collapse or demolition of a supporting structure, rather than to overstressing.

CHAPTER FOUR

Materials

and Environment

in Rome

Do not do unto others as you would they should do unto you. Their tastes may not be the same.

GEORGE BERNARD SHAW

Maxims for Revolutionaries

We shall now consider some of Vitruvius' comments on building materials and climate and have a look at Roman water supply, sewage disposal, and their under-floor system of heating.

4.1 QUALITY CONTROL OF MATERIALS

Vitruvius devoted a great deal of space to the description of suitable materials. The practice of specifying that stone from specific quarries or timber of a particular species be used was continued well into the twentieth century. It ensured that the material had been tested by long experience, for there is no other certain way of attesting durability. He did not, however, advocate transporting materials over long distances. When a suitable quarry did not lie close enough to the site of construction, the durability could be tested by exposure to the weather:

There are several quarries called Anician in the territory of Tarquinii, the stone being the colour of peperino. The principal workshops lie round the lake of Bolsena and in the prefecture of Statonia. This stone has innumerable good qualities. Neither the season of frost nor exposure to fire can harm it, but it remains solid and lasts to a great age, because there is only a little fire and air in its natural composition, a moderate amount of moisture, and a great deal of the earthy. 'Hence its structure is of close texture and solid, and so it cannot be injured by the weather or the force of fire.

This may best be seen from monuments in the neighbourhood of the town of Ferento which are made of stone from these quarries. Among them are large statues exceedingly well made, images of smaller size, and flowers and acanthus leaves gracefully carved. Old as these are, they look as fresh as if they were only just finished. . . . If these quarries were only near Rome, all our buildings might well be constructed from the products of these workshops.

But since, on account of the proximity of the stone-quarries of Grotta Rossa, Palla, and others that are nearest to the city, necessity drives us to make use of their products, we must proceed as follows, if we wish our work to be finished without flaws. Let the stone be taken from the quarry two years before building is to begin, and not in winter but in summer. Then let it lie exposed in an open place. Such stone as has been damaged by the two years of exposure should be used in the foundations. The rest, which remains unhurt, has passed the test of nature and will endure in those parts of the building which are above ground. (Ref. 2.3 Book 2, Chapter 7, p. 50).

A modern accelerated weathering test includes cycles of freezing and thawing and exposure to sunlight and rain (Ref. 1.3, Section 9.10). Vitruvius recommended a similar procedure without the acceleration for which he lacked the technology. We cannot agree with the chemical explanations in terms of the four elements of fire, water, earth, and air, but they would still have been acceptable in the seventeenth and perhaps even the eighteenth centuries.

The commentary on timber suffers from the same defect but the conclusions are sound:

To begin with fir: it contains a great deal of air and fire, very little moisture and the earthy, so that, as its natural properties are of the lighter class, it is not heavy. Hence, its consistence being naturally stiff, it does not easily bend under the load, and keeps its straightness when used in the framework. But it contains so much heat that it generates and encourages decay, which spoils it; it also kindles fire quickly because of the air in its body, which is so open that it takes in fire and gives out a great flame. . . .

Oak, on the other hand, having enough to spare of the earthy among its elements, and containing but little moisture, air and fire, lasts for an unlimited period when buried in underground structures. It follows that when exposed to moisture, as its texture is not loose and porous, it cannot take in liquid on account of its compactness, but, withdrawing from the moisture, it resists it and warps, thus making cracks in the structures in which it is used (Ref. 2.3, Book 2, Chapter 9, p. 60).

4.2 PIGMENTS AND METALS

Vitruvius described in some detail the pigments available in his time. The following mineral colors were found naturally:

Yellow ochre, found in many places, but that from Athens was the best.

Red earth, found in abundance, but the best came from the Black Sea, from Egypt, the Spanish Balearic islands, and the Greek island of Lemnos.

Melian white from the Greek island of Melos.

Green chalk, found in numerous places, but the best came from Smyrna in Asia Minor.

Malachite green from Macedonia.

Armenian blue from Armenia.

An entire chapter is devoted to vermilion, a famous red pigment manufactured from mercury. Another chapter covers the production of black pigment by burning resin; the process was similar to that still used for carbon black. Purple was obtained from a shellfish; the reddest shade came from Rhodes, the darkest from France. The manufacture of blue pigment from copper and the making of brown pigment by burning ochre are also described.

Finally there is a chapter on substitutes. A cheaper purple could be obtained by dyeing chalk with hysginum and the madder root. Malachite green could be copied at less cost by treating blue pigment with a plant called the dyer's weed, and chalk dyed with woad produced a color almost like indigo.

The Romans were equally cosmopolitan in their use of metal. Iron came from Tuscany, Gaul, Spain, and Carinthia. The Indian "wootz" made by repeated hammering and reheating of the iron, which gave good control over the carbon content and produced a steel highly esteemed for swords, was also imported.

Copper came from Cyprus and Asia Minor, tin for alloying with copper to make bronze was brought from Cornwall, and lead was mined in Greece.

The Romans used lead extensively to make pipes by bending the sheet and hammering the ends into a joint. These pipes have been found in various Roman baths (including Bath in England). Molten lead was poured into masonry joints for particularly important structures in place of mortar (Section 5.2) and also was used to protect iron clamps in masonry from rust.

Copper was used extravagantly. The Pantheon (Fig. 3.20) was originally covered with goldplated tiles of copper, which were removed by the Byzantine emperor Constans II. The ceiling of the portico was supported by bronze girders which Pope Urban VIII replaced in 1625 with wooden substitutes. The bronze from these girders was sufficient to cast eighty cannons, some of which are still in the Castel Sant' Angelo in Rome (built as the tomb of the Roman emperor Hadrian).

4.3 WATER SUPPLY AND SANITATION

The Romans did not invent water supply or sewage disposal. Many early civilizations presumably had water-supply channels, but they are not easy to identify today.

Terracotta pipes and a terracotta bath of remarkably modern shape have been found in the Palace of Knossos in Crete: estimated date, 2000 B.C. (Ref. 4.3, p. 7).

King Sennacherib of Assyria built an aqueduct lined with a hydraulic mortar in the seventh century B.C.; King Polycrates' water supply system, which employed a tunnel, in Pergamon, Asia Minor, dates from the second century B.C.

The Romans, however, created a water-supply system on an incomparably vaster scale. The first aqueduct in Rome (the Appia) was built by Appius Claudius Crassus Caecus in 312 B.C. almost entirely underground; the last was completed in A.D. 226, making a total length of 560 km (350 miles) for the eleven aqueducts of Rome.

Most of this distance was under ground, but because the Romans lacked an adequate supply of iron or bronze to build pressure pipes the channels had to cross valleys on embankments or arches. Altogether 80 km were above ground, 60 km of which were high enough to require arches (Fig. 4.1). These elegant arches are still conspicuous in the Roman landscape.

The aqueducts required an army of slaves to build and subsequently to maintain. The Romans knew nothing about thermal movement and did not use expansion joints. Consequently the cracked concrete required constant repair (see Section 3.5).

The Romans had no means of purifying their water. Water from the various springs was kept separate and only the cleanest was used for drinking. There was sufficient water for a sewage disposal system, for fountains, and for flooding the Colosseum from time to time for staging mock naval battles.

Pliny the Elder in the first century A.D. described the aqueducts:

But if anyone will note the abundance of water skilfully brought into the city, for public uses, for baths, for public basins, houses, tunnels, suburban gardens and villas; if he will note the high aqueducts required for maintaining the proper elevation; the mountains which had to be pierced for the same reason (see Fig. 4.2), and the valleys it was necessary to fill up; he will conclude that the whole terrestrial orb offers nothing more marvellous (Ref. 4.4, p. 7).

Frontinus, the best known of the prefects of the water supply, in a book written in A.D. 97 entitled *De Aquis Urbis Romae* (About the Water Supply of the City of Rome), preserved in the monastary of Monte Cassino until it was destroyed in World War II, was more outspoken:

With such an array of indispensable structures carrying so many waters, compare, if you will, the idle Pyramids or the useless, though famous, works of the Greeks (Ref. 4.4, p. 17).

4.1

A reconstruction of the Roman aqueducts at the crossing of the Via Latina. The triple conduit contains the Marcia, Tepula, and Julia; the other contains the Claudia and Anio Novis. (From a painting *Wasserleitungen im alten Rom* by Zeno Dietner in the Deutsches Museum, Munich.)

Frontinus calculated the amount of water supplied by the nine aqueducts of his time (Ref. 4.1, p. 16) as 24,360 quinariae, a Roman measure based on the size of the opening of the ducts. Isaac, after quoting various methods of calculation and measurements of the flow of the Marcia, Virgo, and Claudia springs in the nineteenth century, concluded that the supply amounted to 11 m³/sec (250 million U.S. gallons per day), which was about two-thirds of the supply of Greater London in 1954 and about seven and one-half times the supply of Newcastle-upon-Tyne and Gateshead in the same year, when Newcastle and Gateshead had a population of about 350,000, comparable, as far as we know, to that of Rome in the days of Frontinus.

The amount of water required depends on whether taps were installed to shut off unneeded water. Since the aqueducts operated mostly at atmospheric pressure, the Romans possessed the technological skill required to install taps of wood or metal, which would not, of course, have survived to the present time. If there were no taps, as is generally assumed, a great deal of water would have run to waste and the supply might not have been so generous.

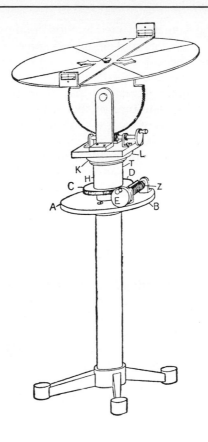

4.2

Dioptra, a surveying instrument described by Hero of Alexandria in the second century B.C. and by Virtruvius in the first. The instrument has graduated vertical and horizontal circles as well as slow-motion screws for both vertical and horizontal rotation. Telescopic sights became possible only in the seventeenth century A.D., but in other respects this instrument is similar to a modern theodolite. We do not know the accuracy of the graduations; however, because the Romans used the dioptra for setting out the tunnels of the aqueducts, which were cut simultaneously from both ends, the accuracy must have been comparable to that of an eighteenth-century instrument.

The aqueducts of Rome were maintained in good condition until the fifth century when the Goths besieged the city and destroyed some of them. Thereafter they gradually deteriorated until the water supply gave out in the eleventh century. Pope Sixtus V in the sixteenth century and Pope Paul V in the seventeenth restored three of the aqueducts which still supply Rome with part of its water (Section 7.9).

Comparable water supply systems on a smaller scale were established all over the Roman empire. The most obvious remains are the arched aqueducts, of which those in Nimes (known as the Pont-du-Gard) and Lyon in France, Segovia, Mérida, and Tarragona in Spain, Constantine in North Africa, and Akko in Israel are well known.

The public baths (thermae) of Rome were open to the general populace either free or for a small entrance fee. They contained cold baths, sometimes hot baths, rooms for exercise and recreation, and heated rooms, somewhat like a modern Turkish bath. The largest of the public baths of Rome, built by the Emperor Caracalla in the early third century B.C. measured 229 x 116 m (750 x 380 ft); it covered more space than the British Museum in London. The plan of the Roman Bath in the English city of Bath (Aquae Sulis) is shown in Fig. 3.21.

Vitruvius, Palladius, and Faventinus wrote about private baths, which were confined to the houses of the wealthy. Plommer considered that the private bath looked rather like a modern bath tub; it was filled through a metal pipe from a metal boiler placed on top of a furnace (Ref. 3.29, p. 13).

There was water-borne sewage before Rome, but on a very limited scale and probably only for important people. A toilet has been found in the palace of Sargon II, an Assyrian king of the seventh century B.C., with a water jug beside each seat. The Palace of Knossos in Crete had a latrine which was cleared by a constantly running brook; this probably dates from about 2000 B.C.

These are, however, isolated examples, compared with the elaborate system built in Rome. Roman *foricae* (toilets), which have been found, were often quite luxurious; the seats were of marble and were usually arranged in groups:

The Roman forica was public in the fullest sense of the term. . . . People met there, conversed and exchanged invitations to dinner without embarrassment. And at the same time it was equipped with superfluities which we forego and decorated with a lavishness we are not wont to spend on such a spot. All round the semicircle or rectangle which it formed, water flowed continuously in little channels, in front of which a score of seats were fixed. The seats were of marble, and the opening was framed by sculptured brackets in the form of dolphins. . . . Above the seats it was not unusual to set niches containing statues of gods or heroes . . . or an altar to Fortune; . . . and not infrequently the room was cheered by the playing of a fountain (Ref. 4.5).

Evidently this would have consumed a good proportion of Rome's water supply, but presumably the Romans had no disinfectants.

It seems likely that every insula (block of flats) had a *forica* and a water supply on the ground floor, and the more important apartments may have had water on higher floors. The sewage was discharged into the River Tiber, mostly through the *Cloaca Maxima*, the great sewer started in the sixth century BC and the only major sewer in Europe earlier than the seventeenth century.

This sewage system was not merely unequalled during the Middle Ages and the Renaissance but was still unsurpassed in the early nineteenth century. Edwin Chadwick in 1842 published *A Report on the Sanitary Condition of the Labouring Population* of Great Britain in which he stated that the sanitary arrangements at the Colosseum in Rome and in the amphitheatre of Verona in the days of the Roman Empire were superior to any facilities existing in London, then the world's largest city. Nobody contradicted him on this point and nobody suggested that conditions were better in any other part of Europe or North America. It is only fair to point out that after the Chadwick Report there was a rapid improvement in sanitation (Ref. 1.3, Section 7.5).

4.4 ACOUSTICS

We know more about Greek acoustics than about any other aspect of Greek building science. We have the texts of many Greek tragedies and comedies and accounts of their performance in other Greek literature. We also have two surviving theaters in Athens (Fig. 4.3) and one in Epidauros; the latter has remarkably fine acoustics.

Our knowledge of the design of both Greek and Roman theaters is based mainly on the account given by Vitruvius in Book 5. The proportional relationships were set out in such detail that it is possible to reconstruct the design (Fig. 4.4).

4.3

The Greek Theater of Dionyssos below the Acropolis in Athens. The original sculptured marble seats for dignitaries are visible in the foreground.

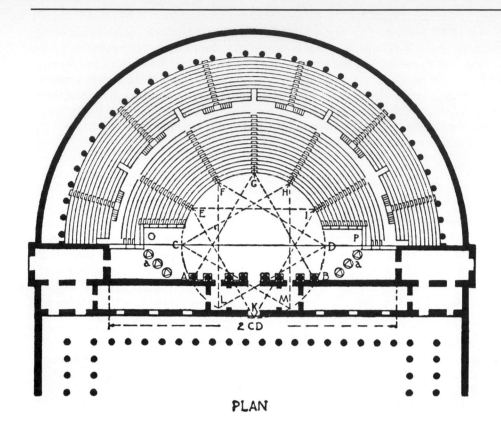

PLAN

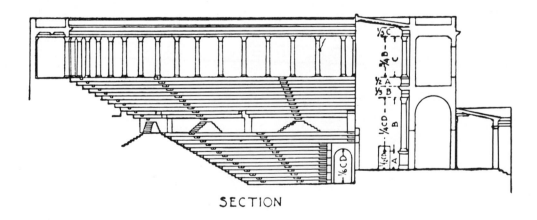

SECTION

4.4

A Roman theater according to Vitruvius (Ref. 2.3, p. 147).

Materials and Environment in Rome

Both Greek and Roman theaters were arranged in a semicircle. The Greek actors performed on a flat space called the *orchestra,* occupying about three-quarters of a circle. This was backed by an open *proskenion* (proscenium) and behind this was the solid *skene*. In the Roman theater the orchestra was reserved for senators and other distinguished spectators and the actors performed on the stage.

The geometric constructions employed by Vitruvius for the two theaters were different. The Greek theater was based on a system of interacting squares, and the Roman theater on a system of interacting triangles. (See also Section 6.3 on the use of triangles and squares in Gothic design.) The acoustic reasons for this geometry are not clear, but Vitruvius' explanations of the nature of sound agree with modern concepts:

Voice is a flowing breath of air, perceptible to the hearing by contact. It moves in an endless number of cirular rounds, like the innumerably increasing circular waves which appear when a stone is thrown into smooth water, and which keep on spreading indefinitely from the centre unless interrupted by narrow limits, or by some obstruction which prevents such waves from reaching their end in due formation. When they are interrupted by obstructions, the first waves, flowing back, break up the formation of those which follow.

In the same manner the voice executes its movements in concentric circles; but while in the case of water the circles move horizontally on a plane surface, the voice not only proceeds horizontally, but also ascends vertically by regular stages. Therefore, as in the case of the waves formed in the water, so it is the case of the voice: the first wave, when there is no obstruction to interrupt it, does not break up the second or the following waves, but they all reach the ears of the lowest and the highest spectators without an echo.

Hence the ancient architects, following in the footsteps of nature, perfected the ascending rows of seats in theatres from their investigations of the ascending voice, and, by means of the canonical theory of the mathematicians and that of the musicians, endeavoured to make every voice uttered on the stage come with greater clearness and sweetness to the ears of the audience. For just as musical instruments are brought to perfection of clearness in the sound of their strings by means of bronze plates or horn *echeia*, so the ancients devised methods of increasing the power of the voice in theatres through the application of harmonics (Ref. 2.3, Book 5, Chapter 4, pp. 138–139).

The design of the echeia was explained in the following chapter:

In accordance with the foregoing investigations on mathematical principles, let bronze vessels be made, proportionate to the size of the theatre, and let them be so fashioned that, when touched, they may produce with one another the notes of the fourth, the fifth, and so on up to the double octave. Then, having constructed niches in between the seats of the theatre, let the vessels be arranged in them, in accordance with musical laws, in such a way that they nowhere touch the wall, but have a clear space all round them and room over their tops. They should be set upside down, and be supported on the side facing the stage by wedges not less than half a foot high. Opposite each niche, apertures should be left in the surface of the seat next below, two feet long and half a foot deep. . . .

Somebody will perhaps say that many theatres are built every year in Rome, and that in them no attention at all is paid to these principles; but he will be in error, from the fact that all our public theatres made of wood contain a great deal of boarding, which must be resonant. This may be observed from the behaviour of those who sing to the lyre, who, when they wish to sing in a higher key, turn towards the folding doors on the stage, and thus by their aid are reinforced with a sound in harmony with the voice. But when theatres are built of solid

4.5

The Roman Theater of Herodes Atticus on the slope of the Acropolis in Athens. It seats 6000 people and is occasionally used for grand opera. The seating is modern. The theater is sunk into the ground on three sides and protected against noise by a heavy *skene* on the fourth side. The motorway visible in the background does not interfere unduly with the performance.

materials like masonry, stone or marble, which cannot be resonant, then the principles of the *echeia* must be applied (Ref. 2.3, Book 5, Chapter 5, pp. 143–145).

The description of the *echeia* is not sufficient to enable us to construct one and none has survived. Nor can we be certain of the nature of the "resonant boarding." Many modern musicians believe in the value of "resonant wood" in a concert hall, but acoustic consultants give no credence to this theory (Ref. 4.2, p. 9).

A number of Roman theaters have survived: in Athens the Theatre of Herodes Atticus on the slope of the Acropolis (Fig. 4.5), in Fiesole near Florence, in Taormina, Sicily, in Lyons and Orange (7000 spectators), France, in Caesaria, Israel, in Aspendus in Asia Minor, in Sabratha in North Africa, and the two theaters excavated at Pompeii. The only surviving theater in Rome, the Theater of Marcellus, was converted into a palace by the Orsini family in the sixteenth century.

In addition there are two surviving amphitheaters in which the seats form a complete oval instead of a semicircle. The Colosseum in Rome (Fig. 4.6) with 87,000 seats (Ref. 4.6, article "Amphitheatre"), which was used as a stone quarry during the

Renaissance, and the smaller amphitheater of Verona which is still used for theatrical performances.

The Greek and Roman theaters were built in the open air presumably because of the difficulty of ventilating large roofed auditoria and evacuating them in case of fire. It would have been impossible in Roman times to roof these spaces without using timber. Acoustical problems are much more difficult when there is no roof to reflect the sound, and some of the theaters were very large. Assuming that the Colosseum was used for spectacles rather than theatrical performances, we still have auditoria with more seats than any modern opera house.

Richardson (4.7, p. 21) thought that the masks habitually worn by Greek actors may have had megaphones of some sort fitted into their mouthpieces. He also suggested that actors would have ascertained the natural frequency of the *echeia* mentioned by Vitruvius, or of the cavities found in the front rows which might have

4.6

The Flavian Amphitheater in Rome, usually called the Colosseum, which had a seating capacity of 87,000. The seats and the stage have been destroyed, for the building has served at various periods as a quarry for building stone (see also Fig. 3.28).

served the same purpose, and then have recited their speeches in a monotone at that frequency. We find it necessary in using the surviving theaters to employ electronic reinforcement or to accept a standard below that of a good modern auditorium. Nevertheless the ancient theaters are remarkably good for their size. As Vitruvius correctly said, the secret of their success lies in the use of steeply raked seats.

4.5   VITRUVIUS ON CLIMATE

In Chapter 4 of Book 6 Vitruvius offered advice on the aspect of various rooms (referring to the northern hemisphere):

Winter dining rooms and bathrooms should have a southwestern exposure, for the reason that they need the evening light, and also because the setting sun, facing them with all its splendour but with abated heat, lends a gentler warmth to that quarter in the evening. Bedrooms and libraries ought to have an eastern exposure, because their purposes require morning light, and also because books in such libraries will not decay. In libraries with southern exposures the books are ruined by worms and dampness, because damp winds come up, which breed and nourish the worms, and destroy the books with mould, by spreading their damp breath over them.

Dining rooms for spring and autumn to the east; for when the windows face that quarter, the sun, as he goes on his career from over against them to the west, leaves such rooms at the proper temperature at the time when it is customary to use them. Summer dining rooms to the north, because that quarter is not, like the others, burning with heat during the solstice, for the reason that it is unexposed to the sun's course, and hence it always keeps cool, and makes the use of the rooms both healthy and agreeable (Ref. 2.3, pp. 180–181).

Vitruvius did not believe in natural ventilation:

Cold winds are disagreeable, hot winds are enervating, moist winds unhealthy . . . . By shutting out the winds from our dwellings we shall not only make the place healthful for people who are well, but also in the case of diseases due perhaps to unfavourable situations elsewhere, the patients, who in other healthy places might be cured by a different form of treatment, will here be cured more quickly by the mildness that comes from shutting out of the winds (Ref. 2.3, Book 1, Chapter 6, pp. 24–25).

Regrettably Vitruvius displayed the kind of chauvinism that is still often expressed today:

But although southern nations have the keenest wits, and are infinitely clever in forming schemes, yet the moment it comes to displaying valour, they succumb because all manliness of spirit is sucked out of them by the sun. On the other hand, men born in cold countries are indeed readier to meet the shock of arms with great courage and without timidity, but their wits are so slow that they will rush to the charge inconsiderately and inexpertly, thus defeating their own devices. Such being nature's arrangement of the universe, and all these nations being allotted temperaments which are lacking in due moderation, the truly perfect territory, situated under the middle of the heaven, and having on each side the entire extent of the world and its countries, is that which is occupied by the Roman people (Ref. 2.3, Book 6, Chapter 1, p. 173).

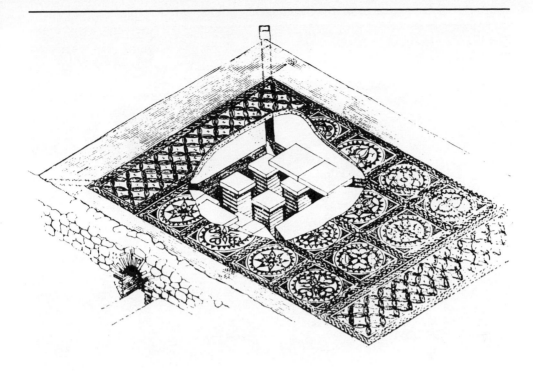

4.7

Sketch of the hypocaust heating beneath a mosaic floor in the Roman city of Verulamium, now St. Albans (Ref. 4.8).

4.6 THE HYPOCAUST

The Romans invented a most ingenious heating system which they used for the hot rooms in baths and for heating some houses. It consisted of a tile floor raised on short brick pillars (Fig. 4.7) and heated by smoke and hot air from a slow-burning furnace. The gases were passed under it and exhausted into the outside air.

This heating system has been found in the ruins of country houses (*villae*) built in the first century B.C. for heating bathrooms and was still a new technique when Vitruvius described it:

The hanging floors of the hot bath rooms are to be constructed as follows. First the surface of the ground should be laid with tiles a foot and a half square (*tegulae sesquipedales*), sloping towards the furnace (*hypocaustum*) in such a way that, if a ball is thrown in it cannot stop inside but must return of itself to the furnace room; thus the heat of the fire will more readily spread under the hanging flooring. Upon them, pillars made of eight-digit* bricks (*laterculi bessales*) are built, and set at such distance apart that two-foot* tiles (*tegulae bipedales*) may be used to cover them. These pillars should be two feet in height, laid with clay mixed with hair, and covered on top with the two-foot tiles which support the floor (Ref. 2.3, Book 5, Chapter 10, p. 157).

* Eight digits equal 148 mm (6 in). Two Roman feet equal 591 mm (1 ft 11 in.).

4.8

Brick columns supporting the hypocaust excavated at Aquae Sulis, now Bath.

Faventinus, writing almost three centuries later, had a similar prescription. The pillars were still made with bessales (eight-digit bricks), but rounded and 2½ to 3 ft high to allow more space (Ref. 3.29, p. 15). The floor was still formed with bipedales (2-ft bricks).

By that time many public baths had been built and hypocausts were used in the tepidaria (tepid rooms) and calidaria (hot rooms). Hypocausts were also used to heat the living room floors in the houses of the wealthy in the colder regions, several of which have been discovered in Britain; excavated hypocausts are exhibited in St. Albans (Fig. 4.7), Bath (Fig. 4.8), and Caerwent in Monmouthshire. It seems likely that a wealthy citizen of Britain had better heating in the third century A.D. than at any subsequent time until the seventeenth century.

It is understandable that the fall of the Roman Empire resulted in the destruction of the baths and the water-supply system because they needed constant maintenance and an organized community effort. The disappearance of the hypocaust is more puzzling, for it required no particular skill to build and little maintenance; furthermore, the withdrawal of the Roman garrisons from Britain, France, and Germany was a gradual process that allowed ample time for its adjustment to a more modest standard of building.

CHAPTER FIVE

The
—
Middle Ages

What is the good of Mercator's North Poles and
> *Equators,*
Tropics, Zones and Meridian Lines?"
So the Bellman would cry; and the crew would reply
"They are merely conventional signs!
Other maps are such shapes, with their islands and
> *capes!*
But we've got our brave Captain to thank"
(So the crew would protest) "that he's brought us
> *the best . . .*
A perfect and absolute blank!"

LEWIS CARROLL

The Hunting of the Snark

In this chapter we shall consider the building technology in the period following the fall of the West Roman Empire to the Renaissance (approximately A.D. 450 to 1450), both in western Europe and the Byzantine and Arab worlds. The special problems of Gothic architecture are considered in Chapter 6.

5.1   THE FALL OF THE ROMAN EMPIRE

The decline of the Roman Empire was a gradual process. Deterioration soon set in at the center. As early as A.D. 69 there were four emperors in one year by the will of the army: Galba, Otho, Vitellius, and Vespasian (who subsequently held the throne for ten years); but in spite of the poor quality of many of the emperors Rome succeeded in maintaining a high standard of competence and integrity in its public service. As in the Third and Fourth French Republics, day-to-day government continued even when power at the top was in doubt. Indeed, most of the great architectural and engineering monuments that have survived were built after the demise of the Republic.

The Romans insisted that all their public servants and all who wished to do business with them speak Latin; thus they established Latin as the language of the Empire, and its adoption by the Roman Catholic Church ensured its retention as the international means of communication. Until the seventeenth century most scientific books were published in Latin and most scientists used Latin for international correspondence in the way that English is used at present.

The Romans also established a system of all-weather roads that remained usable for centuries after the fall of the Empire. However, the courier services lapsed, and as the bridges collapsed they were not replaced, although some have survived. In the city of Rome the Pons Fabricius, the Pons Cestius, and the Pons Aelius have been in continuous use since the days of the Roman Empire and there are bridges in other parts of the former Roman territories still serviceable.

No masonry bridge was built again in Europe until the twelfth century. London Bridge was erected between 1176 and 1209, but it required nineteen arches to cross the Thames and gave constant trouble ("London Bridge is falling down, falling down . . ."), although it lasted until 1820. The bridge over the Rhone at Avignon, also celebrated in song, was built between 1178 and 1186; four of its twenty-one arches are still standing.

The Romans built fortifications along the Rhine and Danube, and long walls were erected where there was no river to protect the frontiers. Toward the end of the fourth century the barbarians (a term used to describe nomadic tribes who could not speak Latin) broke through the defenses, and within a century the Western Empire had disintegrated. Some of these tribes (particularly the Vandals and the Huns) acquired a notorious reputation, but all had a tendency to plunder and move on rather than settle down. Local populaces frequently destroyed the Roman roads to delay the progress of the invaders and this further contributed to the breakdown of communications. The movements of the tribes were prodigious, even by modern standards. The Normans moved from Scandinavia to Sicily and the Vandals, from Germany to North Africa. It took several centuries for a new order to establish itself, but in the meantime life was precarious and travel without an armed escort, dangerous.

The effort devoted to defense against raiding tribes left little time and capital for building or repairs. Thus the Roman fire-fighting, water-supply, and sewage-disposal systems decayed, and because of a lack of raw materials Roman structures were pillaged for their metal, stone, and brick. New masonry buildings were of modest size and primitive construction.

The political recovery started in the eighth century when the kingdom of the Franks, of Germanic descent, grew in power. Charles I (Charlemagne) extended it into Italy, and on Christmas Day 800 was crowned by Pope Leo III "Augustus and Emperor governing the Roman Empire." The title of Roman Emperor was born by a succession of German kings until 1806, when it was abolished by Napoleon I. The Holy Roman Empire, however, was a political power only when the elected emperor had a personal power base, and the kingdoms of France and England soon became more important. By the eleventh century western Europe had acquired the resources to build the great cathedrals that are among the most remarkable structures ever produced.

Although slavery was not formally abolished except in a few free cities, it virtually disappeared with the rise of Christianity until reestablished in the fifteenth century after the first voyages to West Africa. It has been argued that the common man in medieval Europe was worse off than most Roman slaves; but he could not be sold and he could not easily be moved permanently from his home (see Section 6.3). The medieval cathedrals, therefore, had to be built with a comparatively small workforce, and this may account for the long time it took to build them and for the much greater economy of material. Structural safety was not normally an issue in Roman construction, but in Gothic cathedrals collapses were not uncommon.

European construction after the eleventh century was not a revival of Roman architecture but a much more daring form. The revival of Roman ideas during the Renaissance (see Section 7.1) did not completely reverse this trend.

On the environmental side, however, neither medieval nor Renaissance Europe recovered the Roman standards. From the thirteenth century onward the water supply improved slightly, but it remained inferior to that of Rome until the nineteenth century. As the medieval cities grew, the increase in problems of sewage disposal caused appalling epidemics of the plague. The roads also remained in poor condition until the eighteenth century when France founded the Corps des Ingénieurs des Ponts et Chaussées and began building military roads.

5.2 BYZANTINE AND MUSLIM DOMES

While Europe passed through the Dark Ages, the division of the Roman Empire gave the East a new lease of life. Diocletian had abandoned Rome as the capital city at the end of the third century and moved the government to Nicomedia (Izmit, about 50 miles from Istanbul). When the western part of the empire became semi-independent, Mediolanum (Milan), and later, for a short time, Augusta Trevirorum (now Trier, Germany), became its capital. In 330 Constantine I, converted to Christianity, chose Byzantium (now Istanbul), a small city dating from the seventh century B.C., as the new Rome, renaming it Constantinople. In 395 the Empire was partitioned, and eighty years later Rome succumbed to the barbarians. The Byzantine

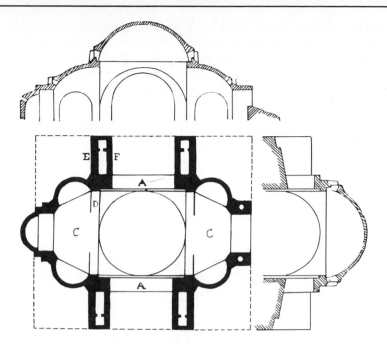

5.1

Plan and cross section of S. Sophia, Constantinople (after A. Choisy: *Histoire de l'Architecture*, Paris, 1899; Ref. 5.16, p. 239).

Empire continued as a great although declining power. Constantinople developed a vital architecture that differed distinctively from that of ancient Rome.

We noted in Section 3.9 that the hoop tension of a dome can be greatly reduced, or even eliminated, by making it shallow. This became the characteristic of the Byzantine dome, which, symbolically, represented the dome of heaven (Ref. 5.7), frequently with a huge figure of Christ painted on its surface (see also Section 7.2).

The first Byzantine dome is also the greatest ever built—the Church of the Holy Wisdom (S. Sophia or Hagia Sophia) in Constantinople (Fig. 5.1). It was erected in the reign of the Emperor Justinian by two architects, Anthemios and Isidorus, in a mere six years and completed in 537. Since the fall of the Roman Empire was reckoned by Gibbon (Ref. 3.6) as the fall of Constantinople to the Turks in 1453, this date is much nearer the beginning than the end of the Roman Empire. It is, however, significantly later than the structures we discussed in Chapter 3.

A description of the construction of S. Sophia appears in *Buildings,* published by the historian Procopius about 560 (Ref. 5.10, pp. 11–31). Procopius also wrote a major treatise on the *Wars* during Justinian's reign, which contained veiled criticisms of the emperor, and a *Secret History,* written about 550, but not intended to be published during his lifetime, which Dewing, the translator of Procopius' works, described as libelous to Justinian (Ref. 5.10, p. 9). In *Buildings* Procopius flatters Justinian and ascribed to him abilities he probably did not have:

So the church of Constantinople, which men are accustomed to call the Great Church, speaking concisely and merely running over the details with the finger-tips, as it were, and mentioning with a fleeting word only the most notable features, was constructed in such a manner by the Emperor Justinian. But it was not with money alone that the Emperor built it, but also with labour of the mind and with the other powers of the soul, as I shall straightaway show. One of the arches I have just mentioned, called *lori* by the masterbuilders (*mechanopoioi*), the one which stands toward the east, had already been built up from either side, but it had not yet been wholly completed in the middle, and was still waiting. And the piers (*pessoi*) above which the structure was being built, unable to carry the mass which bore down upon them, somehow or other suddenly began to crack, and they seemed on the point of collapsing. So Anthemios and Isodorus, terrified at what had happened, carried the matter to the Emperor, having come to have no hope in their technical skill. And straightaway the Emperor, impelled by I know not what, but I suppose by God, for he is not himself a masterbuilder (*mechanikos*), commanded them to carry the curve of this arch to its final completion. "For when it rests upon itself", he said, "it will no longer need the props (*pessoi*) beneath it." And if this story were without witness, I am well aware that it would have seemed a piece of flattery and altogether incredible; but since there are available many witnesses of what then took place, we need not hesitate to proceed with the remainder of the story. So the artisans (*technitai*) carried out his instructions, and the whole arch then hung secure, sealing by experiment the truth of his idea. Thus, then, was this arch completed (Ref. 5.10, pp. 29–31).

Procopius gave an accurate description of S. Sophia, but because the building is fortunately still in good condition it has added little to our knowledge of the structure. He gave no reasons for the design of the church in this unusual and original manner, and his account of the two construction failures dealt mainly with Justinian's role in saving the building, which was probably flattery, notwithstanding the last sentences quoted. The masterbuilders would naturally have consulted the emperor on a matter of such importance if only to cover themselves in the event of failure.

The use of the terms *mechanikos* and *mechanopoios* in place of the traditional *architekton* has caused some writers on Byzantine architecture to suggest that Anthemios and Isodorus might have had been trained in structural mechanics. However, the Byzantine knowledge of mechanics was limited to simple machines (Section 2.5) and would have been of little help in designing a building. Dewing's translation of the terms as *masterbuilder* seems the appropriate one.

Procopius stated that lead was used as the cement that bound the stones together, a practice already employed in ancient Rome (Section 4.2), although not on the same scale:

The piers which I have just mentioned are not constructed in the same way as other structures, but in the following manner. The courses of stone were laid down so as to form a four-cornered shape, the stones being rough by nature, but worked smooth. . . . These were held together neither by lime, nor by asphalt, the material which was the pride of Semiramis in Babylon, nor by any other such thing but by lead (*molibdos*) poured into the interstices (*telma*), which flowed about everywhere in the spaces between stones, and hardened in the joints (*harmonia*), binding them to each other (Ref. 5.10, pp. 23–25).

The dome of S. Sophia subtends an angle of approximately 143° at the center of curvature (Fig. 3.31*b*) so that only a small hoop tension develops (Ref. 3.9). The horizontal reaction or thrust, however, must be absorbed. In the first place the dome

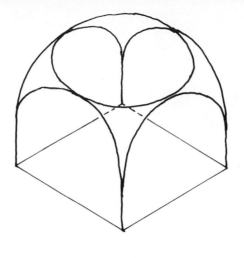

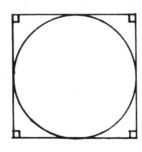

5.2

Conversion of the circular plan of a shallow spherical dome into a square plan by placing it on pendentives, that is, four spherical triangles. Pendentives are not a Byzantine invention. They are found in late Roman Imperial and in Persian domes during the Sassanid dynasty, but these were small structures. In Byzantine and Muslim architecture pendentives became a major structural feature.

is transformed from a round to a square base by placing it on spherical triangles, called *pendentives* (Fig. 5.2). We now have four horizontal thrusts, directed right and left and up and down (Fig. 5.1). The right and left thrusts are taken by semidomes, the other two by massive buttresses.

We therefore have a shallow spherical dome, built from bricks, which is almost entirely in compression; four pendentives transfer the reactions of the dome to four great arches: the horizontal reactions of the dome are then absorbed along one axis by two semidomes and along the other axis by four massive buttresses. This is the system used in subsequent Byzantine and Muslim domes, except that most have four semidomes, two along each axis, sometimes further stiffened by buttresses.

5.3

The Blue (Sultan Ahmet) Mosque built opposite S. Sophia and above the palace of the Emperor Justinian (who ordered the construction of S. Sophia.) The foundation of the palace, recently excavated, may be seen below the mosque. The dome has a diameter of 23.6 m (77 ft). It is supported on massive columns 5 m (17 ft) in diameter and buttressed by four semidomes.

The span of the dome of S. Sophia (Refs. 5.11, 5.16, 5.19, and 5.20) is 33 m (107 ft), considerably less than that of the Pantheon. The actual impression of space is much greater, however, because to this span must be added the semidomes and space between the buttresses, which creates a cross-shaped plan of 76 by 67 m (250 by 220 ft). Furthermore, the dome sits on top of four great pillars, thus rising to a height of 54 m (177 ft) compared with the 43-m sphere of the Pantheon. This tremendous space has been exceeded by a few later buildings. Perhaps only S. Pietro in Rome and the Duomo of Florence (see Sections 7.2 and 7.6) create a greater impression of spaciousness.

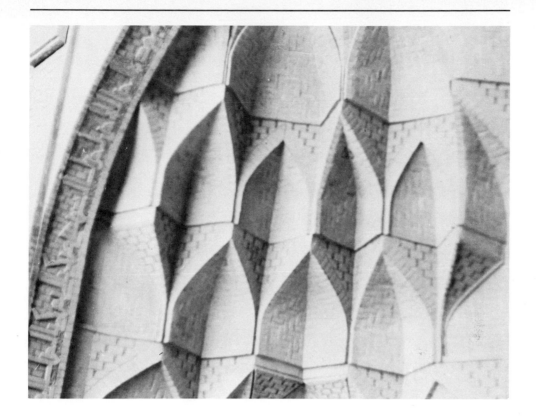

5.4a

The Friday Mosque in Isfahan, Iran, built about A.D. 760. Detail of bond used in the brickwork for the squinches of one of the Iwans (recessed arches).

S. Sophia has been subjected to earthquakes which caused partial collapse of the structure in 557 and again in the tenth century. Other earth tremors produced serious distortions, described, together with the repairs and additional buttressing undertaken, by Mainstone (Ref. 5.11).

The Pantheon, which is in excellent condition, is older by more than 400 years. S. Sophia, however, is a lighter and more daring structure. It is twice as old as Chartres Cathedral (see Section 6.1) and three times as old as S. Pietro (see Section 7.6).

It is appropriate to mention here, briefly, the Muslim and Russian domes derived from S. Sophia.

5.4b

The Great Iwan of the Friday Mosque whose brickwork was later covered with glazed yellow and blue tiles.

Although the Blue Mosque (Fig. 5.3) and the Suleimaniyeh Mosque in Istanbul are exquisitely proportioned and decorated, they do not represent a structural advance, for their domes are actually smaller than that of S. Sophia.

In Persia and the adjoining regions small arches, or squinches, were used as a device for building great arches freehand (Fig. 5.4): the brickwork of each squinch could be laid without supporting formwork; the next squinch was then constructed above it so that formwork was required for only the central portion of an arch or vault (Ref. 5.21). (See also Sections 6.4 and 7.2.) During the late phase of Muslim architecture these squinches became smaller and mainly decorative (Fig. 5.5) and in some buildings, for example, the Alhambra in Granada, pendants were added to make them look like stalactites. This was a degradation (or elevation, depending on one's point of view) of a constructional device to a purely decorative one, which was

The Middle Ages

5.5

Entrance to the Sheikh Lutfullah Mosque in Isfahan, Iran, built in the early seventeenth century. The squinches are much reduced in scale and covered with exquisitely decorated glazed blue tiles.

also characteristic of the last phase of Gothic architecture (Sections 6.5 and 6.6) and roughly contemporary with it.

Another irrational, but often beautiful, structural variation is the onion dome (Fig. 5.6). The swelling of the diameter of the dome provided a visual balance when the dome was small in relation to the mass of the building. Because onion domes were usually of small diameter, they did not present serious structural problems.

Timber domes were probably quite common and may at one time have outnumbered masonry domes. Because timber has good tensile strength, they did not require buttresses or chains. The most notable early example that survives is the Dome of the Rock in Jerusalem (Fig. 5.7). Timber domes were particularly common in Russia where timber was plentiful. The popularity of the onion dome was at least partly due to the relative ease with which the bulbous shape could be built in timber, compared with masonry.

5.6

The Taj Mahal in Agra, India, built from 1632 to 1643 of white marble.
The diameter of the dome is only 18 m (58 ft).

5.3 CATENARY ARCHES AND DOMES

We have pointed out that the circular arch and dome are not structurally the most efficient. A cable supporting its own weight is in pure tension; it assumes a shape called a catenary, whose mathematical equation is $y = c \cosh(x/c)$, where x and y are the horizontal and vertical coordinates and c is a constant. Unless the sag is appreciable, the shape is quite similar to that of a parabola.

If we turn a catenary-shaped structure of uniform thickness upside down, we obtain a structure that is entirely in compression.

Structures approximating the shape of a catenary are found in places in which the building material is so weak that any other shape would collapse. Presumably the shape was obtained by trial and error. Mud huts of approximately catenary shape are found in various parts of the world (Fig. 5.8).

The largest structure of this type is the Great Hall of the Palace of Taq Kisra at Ctesiphon, then the capital of Persia. It was built by Khosrau I, one of the kings of the Sassanid dynasty, about A.D. 550. The Great Hall is a catenary-shaped vault of brick, 34 m (112 ft) high and spanning 25 m (84 ft). The walls at the base are 7 m (23 ft)

The Middle Ages

5.7a

The Dome of the Rock on Mount Moriah, Jerusalem, originally built in A.D. 691. The present dome, apart from minor repairs, is believed to have been built in A.D. 1022 as an exact copy; this would make it the oldest monument of Muslim architecture surviving in its original form. View of the present wooden dome, 20.4 m (67 ft) diameter, covered with gilded lead. Note the pointed arches in the screen wall.

5.7b

The Dome of the Rock is reported to be a copy of a similar wooden dome over the Holy Sepulchre in Jerusalem which has long disappeared (Ref. 5.7). Section through the double wooden dome, according to C. J. M. de Vogüé, *La Temple de Jerusalem*, Paris, 1865 (Ref. 5.7, plate 37).

thick. The vault is built of brick, laid in horizontal courses, but the weight of the material is probably sufficient to cause it to act like a catenary arch (see Section 3.9). After Ctesiphon was conquered by the Muslims it was turned into a mosque and when the capital was moved to Baghdad it became a quarry for bricks. The ruins are still standing in what is now a suburb of Baghdad. The arched shape of the Ctesiphon vault has been the inspiration of modern catenary thin-shell vaults (Ref. 5.1).

The Ctesiphon vault is the largest surviving example of an old Middle Eastern tradition of building arches in brick in the shape most suited to the material (Ref. 5.8, Plates 6–18), and it seems likely that the pointed Saracen arch (see Section 6.1) is derived from this tradition.

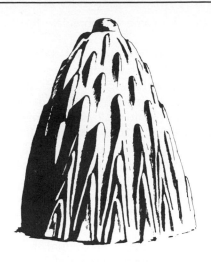

5.8

A traditional mud hut in Central Africa.

5.4 SCIENCE AND TECHNOLOGY IN THE MIDDLE AGES

Following the move to Byzantium Greek replaced Latin as the official language of the Empire and of the Eastern Church. The Greek philosophers continued to be studied and minor additions were made to Greek science and mathematics (see Section 2.2). The Library of Alexandria was destroyed during the Arab conquest in 641, but most of the Greek scientific texts were probably still in existence in Constantinople when the Turks captured it in 1453. Gibbon (Ref. 3.6) stated that 120,000 manuscripts were destroyed when the city was sacked.

The Muslim conquest started in 633, a year after the death of Mohammed. Damascus fell in 635, Tripoli in 643, and Cyprus in 655. Constantinople itself was besieged unsuccessfully in 669. In 711 the Arabs and Berbers captured most of Spain from the Goths. In 732 they were defeated decisively by Charles Martel, the grandfather of Charlemagne, at Poitiers in southern France. In 740 the Byzantine emperor Leo III defeated another Arab force at Akrainos in Asia Minor. The Muslims were halted but only after they had conquered most of the countries that preserved the Greek heritage.

The Nestorian Church, which used the Syriac language, an Aramaic dialect akin to Hebrew, had been persecuted by the Greek Orthodox Church after its doctrine was declared a heresy by the Council of Ephesus in 431. The Nestorians had created an important school at Edessa (now Urfu in Turkey). They did not resist the Arab conquest and were recruited as translators. The Arabs were particularly interested in

Greek medicine, but they soon came to appreciate the value of the mathematical books.

Many of the Greek texts were not translated directly into Arabic but first into Syriac. Many have been lost, for we have references to texts that do not survive and there may be others of which we have never heard. Some were translated directly into Latin, but some of the most important books, notably the Almagest of Ptolemy and the Elements of Euclid (see Section 2.2), have come to us via Arabic. The latter was first translated into Latin about 1120 by Abelard of Bath, who obtained an Arabic copy of Spain to which he had gone disguised as a Muslim student. The Christian reconquest of Spain took from the tenth to the fifteenth centuries, during which time there was much cultural interchange between the various Muslim and Christian kingdoms. After the Christian conquest many Muslims and Jews remained behind and helped with the translations. Some Arab texts were translated into Latin via Hebrew.

The revival of learning in Europe had been gradual and in the twelfth century some of the scholastic guilds began to organize themselves as a *universitas magistrorum et scholarium,* a corporation of teachers and students. Universities were founded in Salerno (southern Italy), Bologna, Paris, and Oxford, and many more followed in the thirteenth and fourteenth centuries. These universities sought legal recognition to confer the degrees of magister (master) and doctor (teacher) and eventually applied for international recognition from the pope or the Holy Roman Emperor.

For the purpose of recognition the universities had to demonstrate their ability to teach the full range of the *studium generale,* the seven liberal arts that were organized in the trivium (grammar, rhetoric, and logic) and quadrivium (arithmetic, music, geometry, and astronomy). Study of most of the quadrivium became possible only after Latin translations of the ancient Greek books had become available.

After 1400 scholars left Byzantium in increasing numbers, for it was clearly doomed by the encirclement of the Turks, and settled in western cities, particularly Venice, bringing their knowledge with them.

An important technological change was already taking place in early medieval Europe, which lacked the labor of ancient Rome for mining and metalworking. This was partly due to the abandonment of slavery and partly to the loss of productivity resulting from constant petty warfare (or robbery, depending on one's point of view). It is also likely that the slow decay of the Roman water-supply and sewage-disposal systems and the loss of Greek medical knowledge (which survived in the Arab world) reduced the life expectancy.

Mechanical power had been used before the Middle Ages. Vitruvius (Ref. 2.3) described the Archimedian screw (see Section 2.5) and the waterwheel for grinding corn as well-established contrivances, and ancient Alexandria (see Section 2.3) displayed many ingenious mechanical toys. In the Middle Ages, however, these machines multiplied to do the work previously done by people. The Domesday Book, the survey of England made about 1086, listed several thousand waterwheels. The first windmills appeared in the twelfth century, and machines run on horsepower also became common. Few attempts had been made in the ancient World to utilize the traction power of animals, and the modern harness was introduced into Europe from Asia in the Middle Ages. After the renaissance of Roman engineering the quality of all these machines was greatly improved.

5.9

Moat, masonry wall, and defensive tower of the Bishop's Palace in Wells, Somerset, England, built in the early thirteenth century. (*Photograph by Judith Cowan.*)

5.5 THE EMPHASIS ON SECURITY

The period from the withdrawal of the Roman garrisons (in Britain 410) to approximately the year 1000 was one of great insecurity, which is reflected in the small size and the thick walls of the buildings that survive. Settlements were surrounded by walls if they had enough wealth to afford them but they were mostly of timber and few survive. Existing city walls generally date from the period 1000 to 1500 when they were rendered obsolete by the invention of gunpowder and cannons. After 1000 churches became more substantial monuments (the Romanesque or Norman style in England), but the walls were still thick. From the end of the twelfth century there was a marked refinement in structural sizes (the Gothic style). Churches, regarded as final defensive positions in earlier days, were being built with large glass windows. The contrast is illustrated in Wells in the west of England where the cathedral (Fig. 6.13) is a Gothic structure with large windows, but the bishop's castle is surrounded by a moat and thick walls (Fig. 5.9) that date from the early thirteenth century and are contemporary with the cathedral.

The most extraordinary architectural features of this period were the towers found in some Italian cities. Not only were people afraid of their enemies but neither did they trust the other noble families in their own city. The towers were both defensive positions and observation points from which enemy movements could be followed.

Bologna had forty-one towers, of which the most spectacular are the two that lean, La Garisenda and L'Asinelli (Fig. 5.10). Both were built between 1100 and 1109. La Garisenda, originally about 60 m (200 ft) high, was mentioned by Dante in his Inferno and also on an old gravestone, from which sources it has been concluded (Ref. 5.2) that subsidence of the foundation tilted it quite gradually. In the fourteenth century the top of the tower was dismantled and its height is now 48.16 m (158 ft). It is 8 m (26 ft) square at its base and its walls are 2.35 m (7 ft 9 in.) thick at that level. It is now 2.32 m (7 ft 7 in.) out of plumb, which is about half the lean of the Tower of Pisa. It seems more, however, because it is only 11 m (36 ft) from the taller tower (Fig. 5.11) which has a height of 98 m (322 ft), and a lean of about 1.2 m (4 ft).

It is perhaps appropriate to mention here that many medieval buildings had defective foundations, which were rare among the surviving buildings of ancient Rome, Greece, and Egypt. One does not normally dig up foundations without good reason and we are therefore much better informed about the superstructures than about the foundations. Our knowledge of the latter is based mainly on visible failures. A startling example is Winchester Cathedral (Fig. 5.12), the longest of the medieval cathedrals, which has massive walls 24 m (80 ft) high. The foundations were only 3 m (10 ft) below ground level and consisted of whole beech logs overlaid by a weak concrete (not comparable in strength to Roman or modern concrete). There was no significant enlargement of the width to spread the load of the walls. The subsoil contained a thick layer of peat. It is surprising that these foundations gave little trouble until 1905; perhaps the decay of the logs was due to a lowering of the water table resulting from the activities of the growing City of Winchester, which brought the previously submerged logs above the water table. The consulting engineer, Sir Francis Fox, decided to excavate to a solid foundation and underpin the walls with concrete. The repairs were started in 1906 and the inflow of water was so great that it proved necessary to employ a diver to do the work. The nature of the subsoil encountered is shown in Fig. 5.12.

There are other examples of elaborate foundations, however. Viollet-le-Duc examined Amiens Cathedral in the nineteenth century and found a bed of artificially compacted clay about 0.4 m (15 in.) thick, a bed of concrete (which was probably not very strong) 0.4 m thick, fourteen courses of medium quality stone, each approximately 0.3 m thick, and then three more courses of hard sandstone. This foundation was surrounded by large blocks of rubble.

5.10

The two leaning towers of Bologna. La Garisenda is in front and l'Asinelli is behind. The taller tower is 98 m (322 ft) high. The lower tower has a lean of 1:20 (3°).

The Middle Ages

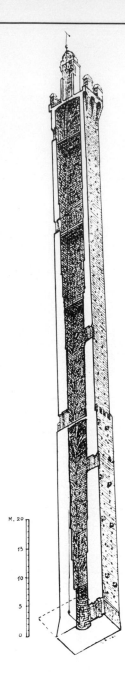

M. 20

15

10

5

0

5.11

Section through the Asinelli Tower in Bologna. The lower walls are thick, but there is a substantial amount of usable room near the top of the tower. (Ref. 5.2, p. 16).

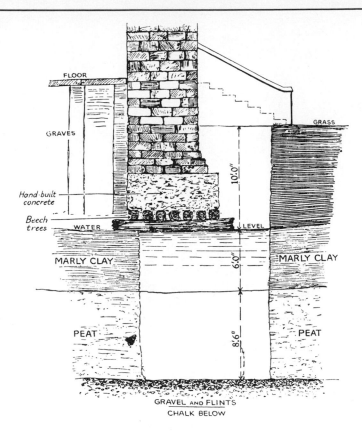

FLOOR

GRAVES

Hand-built
concrete

Beech
trees

WATER   LEVEL

10'0"

GRASS

MARLY CLAY

6'0"

MARLY CLAY

PEAT

8'6"

PEAT

GRAVEL AND FLINTS
CHALK BELOW

5.12

The foundations of Winchester Cathedral as they were found by a diver in 1906 [*J. RIBA,* Vol. XV, third series (1907–1908)].

5.6   MEDIEVAL TIMBER STRUCTURES

A few European timber structures that are known to be medieval survive. The best known is the timber roof of the Great Hall of the old Palace of Westminster, built in the reign of Richard II by his master carpenter, Hugh Herland, from 1394 to 1402. In addition, many small buildings throughout western Europe may have medieval timber structures, although few have been authenticated as being earlier than 1400. The vernacular forms of timber construction changed little between the eleventh and seventeenth centuries, and some therefore are likely to be older than 1400.

The weakness of the vernacular European timber structure was in the joint that generally required halving the timber. The principal types of medieval timber joint were the mortise and tenon and the dovetail (Fig. 5.13). Pegs of hardwood were more common than nails because medieval iron nails were rather soft, rusted without repeated varnishing, and were handmade and quite expensive.

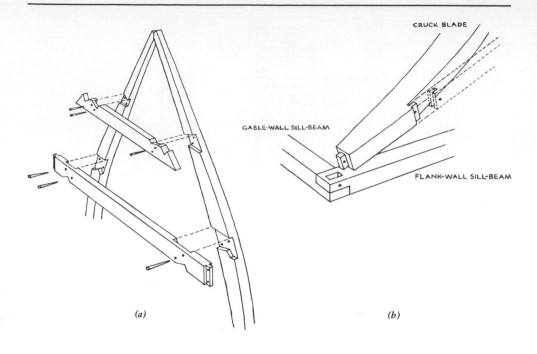

CRUCK BLADE

GABLE-WALL SILL-BEAM

FLANK-WALL SILL-BEAM

(a)

(b)

5.13

Medieval timber joints
(a)   Dovetailed joint in a cruck frame.
(b)   Mortise-and-tenon joints (Ref. 5.17).

In some old timber buildings a strong, if crude, joint was made by the simple expedient of terminating a timber vertical at a branch, placing the horizontal in the Y-shaped crutch, and fixing it by tying.

Timber was also used as wall-framing material, usually without diagonals. The spaces between were filled with brick or more commonly with wattle and daub; that is, thin pieces of wood were joined in a basket-weave pattern, covered with clay, sometimes with an admixture of dung and horsehair, and finished with a coat of plaster or whitewash (see also Section 8.7).

Timber posts were sometimes combined with the roof by choosing suitably curved timber (Fig. 5.13a) called crucks.

Roof trusses did not always have diagonals (Ref. 5.9, pp. 114–120) and in this respect it would seem that medieval timber structures were less advanced than those of Rome (see Section 3.4).

The simplest form after the cruck roof is one with a king or crown post and a collar beam (Fig. 5.14). If two symmetrical posts were used, they were called queen posts and could stand alone or with a crown post. Verticals below the collar beam were called hammer posts, and the horizontals on which they rested were hammer beams. The hammer posts could be used with or without verticals above the collar beam. A

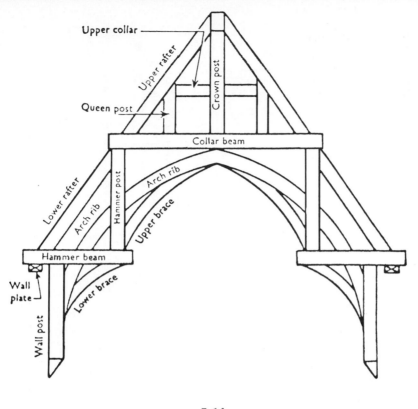

5.14

Nomenclature of the members of medieval timber trusses (Ref. 5.18).

roof of the latter type is illustrated by Villard in a notebook believed to date from the mid-thirteenth century (Ref. 1.2). Curved ribs could be added to stiffen the hammer posts. There are no authenticated medieval timber structures that employed diagonal members in the Roman or modern manner.

The roof of Westminster Hall, mentioned in the opening paragraph of this section, is shown in Fig. 5.15 and diagrammatically in Fig. 5.14. It has been analyzed on a number of occasions, most recently by Heyman (Ref. 5.18). The roof has a span of 21 m (69 ft) and a slope of 55°, corresponding approximately to that of a Pythagorean triangle with sides 3:4:5, which may have been used for setting out the truss. One factor in the design of the roof was the difficulty of obtaining large timbers longer than about 12 m (40 ft), so that the rafters had to be made from two pieces. Heyman concluded that the vertical loads were taken entirely by the rafters, the collar beam, and the crown post; the wind loads were resisted by the rafters, the collar beam, and the arch ribs, which were partly in tension. Heyman thought that the hammer beams, hammer posts, queen posts, and curved braces were virtually unstressed in this truss.

Traditional Chinese and Japanese timber roofs employed an entirely different type of roof framing. They did not have rafters. Instead the purlins were carried on a layer of collar beams (Fig. 5.16). A Chinese roof of this type of the Fo Kuang Temple in

5.15

The hammer beam roof of Westminster Hall, built from 1394 to 1402 by Hugh Herland for Richard II's Palace of Westminster (Ref. 5.5).

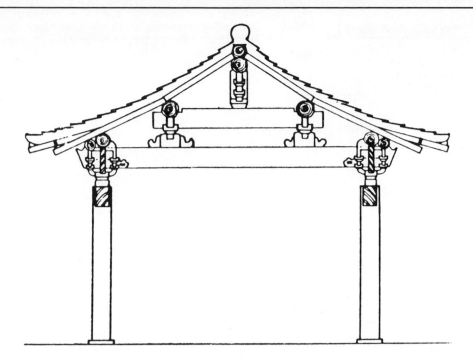

5.16

Framing of Chinese and Japanese traditional timber roofs.

Shansi Province (Ref. 3.19, p. 90) is reputed to date from A.D. 857, and similar roofs were still being built in the twentieth century. Ancient Japanese roofs are rare because of the Japanese custom of replacing the timber structures of important buildings at regular intervals with new ones of identical size and construction. There is no reason to doubt the claim that the Ise Shrine (Ref. 5.4) is identical to one built in 685, but it has in fact been rebuilt fifty-nine times since then. The same applies to other ancient Japanese buildings.

The traditional joints of Chinese and Japanese timber structures also differed from those of Europe. The latter employed shear connectors (Fig. 5.14), whereas the former relied mainly on compression. Thus the Chinese or Japanese joints (Fig. 5.17) provided a substantial bearing and usually an upturned edge to resist horizontal movement; pegs, nails, or binding (Fig. 1.1) were not required, however.

5.17a

Traditional timber joints in temples in Kyoto, Japan. Interior view of beam-column junction.

5.17b

Exterior view of gable end of another temple in Kyoto.

5.7 MEDIEVAL BRICK AND MASONRY STRUCTURES

Brick was a less common material in the Middle Ages than in Roman or modern times. The entire medieval brick-making process was manual. An English statute of 1477 (Ref. 5.12, p. 39) required that the clay be dug before November 1 "stirred and turned" before February 1, and "wrought" after March 1. This was apparently intended to utilize the winter frost and rain for breaking down the clay. It was then puddled into a dough with shovels. The bricks were molded by hand into a wooden form without frogs (which were an eighteenth century invention) and fired with wood or, when available, coal. Because of the amount of labor required, bricks were not necessarily cheaper than natural stone and were used mainly when stone could not be found; for example, in East Anglia.

Unlike Roman bricks, English bricks were never square, but the dimensions varied from place to place. Most cities, however, had a local standard. A mold was kept at the town hall and fines were imposed for the production of bricks that did not conform. The bricks generally ranged from 11 to 8 in. (280 to 203 mm) in length, slightly less than half that dimension in width, and had a thickness of 1¾ to 2½ in.

5.18

Church built with knapped flints and sandstone quoins.

(44 to 63 mm). National standardization occurred in England in the fifteenth and sixteenth centuries, after which bricks were generally 9 × 4½ × 2 in. (229 × 114 × 51 mm) until the seventeenth century, when their thickness increased to 3 in. (Ref. 5.12). The reason for the thinness of the bricks presumably lay in the difficulty of firing a thicker brick. They apparently distorted on firing and consequently required thick joints. In a medieval building the mortar joints tended to form approximately one-fourth of the volume of the brickwork, whereas a sixth would be high today. The mortar was mixed from sand and lime and was much weaker than Roman or modern mortar. It also had a much slower setting time than modern mortar (Ref. 5.6, Appendix G, p. 262).

In England the most notable medieval brick buildings are Hampton Court near London, started by Cardinal Wolsey in 1514, several Cambridge colleges, and Hurstmanceaux Castle in Sussex, built in the midfifteenth century. In Holland, Denmark, North Germany, and other parts of northern Europe brick became common from the twelfth century on mainly because of a shortage of local stone. The Marienkirche in Lübeck is a typical example of German brick Gothic. In France Albi Cathedral in the southwest, begun in 1282, is also of brick.

In Italy, where the Roman tradition was still strong, brick was even more common. Many notable medieval brick structures such as the Basilica of S. Ambroglio in

5.19

Ruins of Glastonbury Abbey, reputedly the burial place of King Arthur, which decayed after the dissolution of the monasteries by Henry VIII. Note the rubble core exposed at the right-hand side of the photograph. (*Photograph by Judith Cowan.*)

Milan, survive in northern Italy. Brick was also a popular building material in Moorish Spain, and after the Christian reconquest Moorish masons continued to use it.

The most common bond in English medieval brickwork (the English bond) consisted of alternate courses of headers and stretchers. However, Flemish bond, that is, alternate headers and stretchers in each course was another form.

Flint consists of silica nodules of organic origin, believed to be derived from siliceous sponges found in the chalk formations of southern England. Because the chalk was too soft, knapped (split) flints were used as building stones. Although an attractive building material, corners were difficult to make; therefore flints were normally used in conjunction with brick or stone (Fig. 5.18).

Building stones in the Middle Ages were most often cut in quarries nearby, for the poor quality of the roads made overland transport laborious and, for large blocks, impossible. If, however, the building site was near water, the stone could be moved economically from distant quarries also near water. Thus the stones of Canterbury Cathedral were quarried at Caen in French Normandy.

Because of the difficulty of assembling a labor force for transport, the largest individual pieces of stone in medieval construction were much smaller than the largest in Egyptian, Greek, and Roman construction.

Dry construction was rare. In early medieval work the stones were roughly cut and in places the mortar joints were excessively thick. However, the quality improved about 1000. In England Norman stone masonry is generally better than Anglo-Saxon, but in the twelfth century there was further improvement and Gothic stone masonry was usually superbly finished on the outside.

The carefully dressed masonry was used only for the outer facing. In early medieval construction walls were very thick, and in Gothic construction there were still some thick parts; for example, the buttresses, the cores of which consisted of rubble in lime mortar, a medieval type of concrete (Fig. 5.19). This construction at first sight resembles the Roman *opus caementitium* (see Sections 3.5 and 3.6) which consisted of masonry formwork with blocks of stone as aggregate and a cast cement mortar. There were two important differences, however. In the Middle Ages the rubble core was probably laid in the modern manner; that is, the mortar was placed in layers with the stones on top, not cast over the stones. Medieval lime mortar was weak by comparison with Roman cement mortar; thus the strength of the wall depended more on the dressed stone outside than on the rubble core.

The finish of stone blocks was often confined to the visible outside, and the bearing between the individual blocks was not so good as in modern masonry. In some cathedrals, notably in Italy where marble was frequently used for the facing but never for the core, stones with widely differing moduli of elasticity deformed differentially when they were loaded (see the footnote in Section 6.5) and when the temperature changed. Many reports have told of medieval masonry cracked and out of plumb by several centimeters. This condition has been observed in particular in the piers under the towers of the taller cathedrals and is not necessarily dangerous (e.g., see Ref. 5.13, discussion). Some cathedrals appear to have survived for centuries with cracked and nonvertical piers.

5.8 MEDIEVAL GLASS

The use of glass as a building material was rare in the ancient world, although a few glass panes have been found in windows in Pompeii and the Middle East. Window glass was not common in Rome, and Vitruvius, who dealt meticulously with every building material (see Section 2.1), did not mention it.

In the Middle Ages the use of glass was already becoming common. The stained glass windows of the Gothic cathedrals are not merely great works of art; they must also have presented formidable structural problems.

The two traditional types are crown and cylinder glass. Crown glass was made by blowing the liquid glass into a rough globe, and then spinning it until by centrifugal force it attained a wheellike shape. The thickness, maximum at the center of the disk, diminished toward the rim. A pane of crown glass of this type dating from the fifth century A.D. has been discovered in Jordan in the Middle East. From there the process was brought to Venice and in medieval times crown glass panes were used throughout Italy. Because of its variable thickness and small round size, medieval crown glass had limited use, but later, after improvement, the process became important (see Section 7.8).

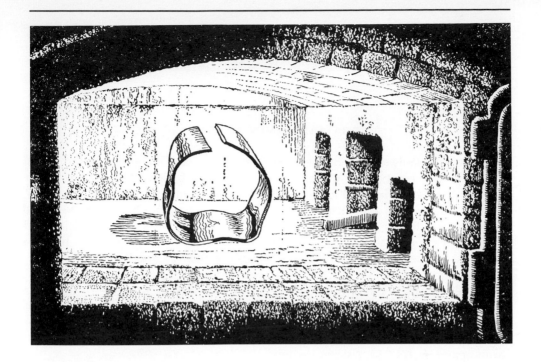

5.20

A cylinder of glass split longitudinally and ready to be opened into a sheet
by gentle heat in an oven (Ref. 5.14).

Cylinder glass was also made in Roman times by a process described by a
medieval author, the Benedictine monk Theophilus, in a manuscript written in the
tenth century (Ref. 5.14). In this process the liquid glass was blown into a broad bulb
which was given a cylindrical form by swinging. One end of the bulb was pierced
and the aperture widened to equal the bulb's greatest diameter. The same was then
done to the other end. Next the cylinder was annealed and split longitudinally with a
red-hot iron, reheated on the flat bed of an oven (Fig. 5.20) on which it opened out
on softening and could be flattened into a sheet with iron tongs and a smooth piece
of wood.

Medieval glass had many imperfections, small air bubbles in particular, and much
of it was translucent rather than transparent. These bubbles, however, gave a bril-
liance to the colors of the stained glass and had to be artificially introduced into
modern "antique" glass (Ref. 1.3, Section 9.3). The techniques for staining glass
were already described by Theophilus in the tenth century. The individual pieces of
glass were usually small and were assembled into larger units with lead cames or
sometimes still larger units with iron bars protected from rusting by varnish. These
assemblies of glass and metal were then attached to stone mullions and transoms.

Window glass came into common use in churches in the twelfth century and in
palaces in the fourteenth. By the beginning of the sixteenth century glass was a
normal building material in middle-class houses.

An early example of the domestic use of glass is in a toilet in the Palace of Westminster; in 1238 Henry III ordered the insertion of glass into the iron-barred window "so that the chamber may not be so draughty as it has been" (Ref. 5.14).

5.9 ENVIRONMENTAL ASPECTS

Medieval standards of hygiene fell far below those of ancient Rome. In the city of Rome all the ancient aqueducts had ceased to function by the eleventh century and the supply was not restored until the sixteenth (see Section 4.3).

In Paris the Roman aqueduct built in the reign of the Emperor Julian was destroyed by Norse invaders in the ninth century, and for three hundred years the Seine and wells were the only water supply. In the twelfth century two monasteries built an aqueduct for their own use. This structure was eventually turned over to the city.

In London the River Thames and wells were the only sources of water until 1237 when Gilbert de Sandford made an agreement with the city to build a water pipe from a spring on his estate at Tyburn (where the Marble Arch now stands). Private households collected their own water from the river, public fountains, or wells or bought it from a water carrier. In London at the end of the fifteenth century there were approximately 4000 water carriers.

The danger in all medieval water sources was the pollution from cesspools and, to a lesser extent, cemeteries. Frontinus' book (see Section 4.3) was known in some monasteries, and several monastic pipelines were built to convey water from a clean spring. Lead water pipes were still used in the Middle Ages, but lead was becoming scarce and wooden pipes began to appear. The hole was laboriously cut with an auger through the solid log.

Cleanliness, which had been highly regarded by the Greeks and Romans, was less important to the early Christians. Moreover, the medieval world was hampered by the inadequacy of the water supply.

Hot public baths were reintroduced into Europe in the thirteenth century after the Crusaders had learned to appreciate the bathing establishments of the Muslims, but they gradually disappeared in the fifteenth century because of objections to communal bathing in the nude. Because water was scarce, it had to be used by many people without change, and the effect of medieval baths on public health may not have been wholly beneficial.

By the thirteenth century, however, the supply of soap improved with consequent effect on hygiene. The Greeks had cleaned themselves by rubbing with oil and then removing the dirty oil with a scraper. Soap made from plant ashes containing alkali and either tallow or olive oil was invented by the Celts in France in the second century A.D. and used in the late Roman period for washing clothes.

Sewage disposal was reasonably effective in small communities. Castles normally had the topmost parapet supported on corbels or brackets projecting a few meters beyond the wall. Early toilets were built in a corner and discharged straight into the moat where aquatic plants and fishes performed the purification. The system worked well as long as the population of the castle was not too large. Later toilet towers provided facilities at each level which discharged into the moat through a drain.

In the late Middle Ages many episcopal and royal palaces had brick sewers that discharged into a river. Sewers of this type were built for the Palace of Westminster and Hampton Court, both of which emptied into the Thames. Brick sewers have also been found at Nonesuch and Eltham Palaces. At Hampton Court the sewer was big enough for a man to enter [5 ft (1.5 m) in diameter] and was required by Cardinal Wolsey to be cleaned out at intervals. In the Middle Ages the Court moved on when the smell became intolerable, and in the reign of Henry VIII it was still laid down in household ordinances that the Court be moved from time to time so that the palace could be "cleansed and sweetened" (Ref. 5.12, p. 78).

In the big cities, however, hygiene was a great problem. In 1395 a Paris ordinance forbade the throwing of nightsoil out of the windows of upper stories, but apparently the practice persisted. In the early fifteenth century some of the open roadside ditches which served Paris as sewers were covered. In the middle of the sixteenth century Henry II asked the French Parliament for money to build the city a proper sewer but the suggestion was made that it should come out of the royal revenue and nothing was done until the seventeenth century. London had to wait even longer (Ref. 1.3, Section 7.5).

Epidemics were frequent and virulent. The Black Death, probably bubonic plague, which raged through Europe between 1347 and 1351, wiped out entire villages. It is estimated that a third of the population died from it. There were many epidemics throughout the Middle Ages, generally with very high death tolls, but only in the fifteenth century were ordinances that required cleansing and quarantine enforced. By that time also the Greek medical texts had reached Europe by way of the Arab world and infectious diseases claimed fewer lives.

Hypocaust heating was forgotten after the fall of Rome. Heat came from a large fire in the center of the great hall and the smoke escaped through a hole in the roof, which was protected from wind and rain by a lantern equipped with louvers. Hampton Court still had a central hearth in the sixteenth century.

In multistory buildings, for example, the keep of a castle, it was necessary to build a ground-floor fireplace against a wall, a practice that gradually became more common. Flues and chimneys began to appear in the thirteenth century in England and in the fourteenth in northern Italy.

Heating, however, remained a luxury, and in winter the great hall also served as sleeping quarters for all but a privileged few.

The great cathedrals were unheated and unheatable (see Section 6.3 for Villard's device for keeping the bishop's hands warm), but in summer in southern Europe the thick masonry walls made them pleasantly cool and still provide an agreeable environment on a hot afternoon.

The standard of lighting in the Middle Ages also fell below that of ancient Rome (Ref. 5.15). Numerous multiwick oil lamps have been found in Roman barracks and houses, which suggests that they were available at least to the army and the middle and upper classes. Candles of tallow and beeswax, already described by Pliny, were more popular in medieval times. Beeswax was inedible and had an agreeable smell, but its supply was limited. Tallow and the vegetable or fish oil used in Roman lamps were all edible and thus lighting competed with the supply of food which in the Middle Ages was never plentiful. Candles therefore were used mainly for religious purposes and by the wealthy.

Acoustics apparently received no attention in medieval times. The Greek and Roman drama disappeared from Europe, and medieval plays were frequently performed in the market place. There are no reports of buildings especially designed for drama in medieval times. Theater design, however, became a major preoccupation of Renaissance Italy and Elizabethan England (see Section 7.10).

Gothic cathedrals provide superb auditoria for organ music and may have had great influence on the design of the modern organ in the sixteenth and seventeenth centuries. Organs have been used in Christian churches since the ninth, although the oldest surviving instrument is a small organ in the church in Syon, Switzerland, which dates from the late fourteenth. The organs in the large Gothic cathedrals are all of more recent date. We do not know how the medieval organs sounded. It seems more likely that they were designed for the cathedrals than vice versa.

CHAPTER SIX

The

Gothic Interlude

"If Mr. Ruskin be right", wrote the reviewer soon after the publication of The Stones of Venice *in 1853, "all architects, and all architectural teaching of the last three hundred years, must have been wrong."*

"This is indeed precisely the fact," replied Ruskin in a later edition, "and the very thing I meant to say, which indeed I thought I had said over and over again. I believe the architects of the last three centuries to have been wrong; wrong without exception; wrong totally, and from the foundation. This is exactly the point I have been trying to prove, from the beginning of this work to the end of it."

J. G. LINKS

The Stones of Venice (Ref. 6.25)

We are devoting a separate chapter to Gothic architecture, one of the most remarkable feats of structural ingenuity. Although today we can analyze its structure, its design remains a mystery.

6.1 THE POINTED ARCH

The designers of Gothic buildings did not invent the pointed arch. It had already been freely used by the Arabs in the seventh century, for example, in the Dome of the Rock (Fig. 5.7a) in Jerusalem. The first crusade led to the establishment of the Christian Kingdom of Jerusalem in 1099. Pointed arches were first seen in Europe at Autin Cathedral in eastern France about 1120 and at the Abbey of St. Denis near Paris about 1140. This would have allowed time for Crusaders returning from Jerusalem to digest the new ideas of the East; it has been reported as well that Saracen prisoners with knowledge of construction were brought back to Europe (Ref. 6.1, p. 28).

Another theory, due to Professor K. J. Conant, gives priority to pointed arches in Italy about 1070, at the Monastery of Monte Cassino, which allegedly were built by craftsmen from Amalfi. From the ninth to the twelfth centuries Amalfi was a free city in competition with Venice and a naval power that traded with the Saracens.

Sir Christopher Wren in the seventeenth century considered that the origins of Gothic style were to be found in the Saracen buildings seen by the Crusaders, but this view has been disputed by many architectural critics.

Whether the pointed arch in an age of poor communications was an original idea or an adaptation, with it the Gothic master builders created an entirely new and daring structural form.

The same applies to the second Gothic ingredient, the ribbed cross vault, which was first used in northern Europe in 1093 in Durham Cathedral. Cross vaults, however, made their appearance in the Basilica of Maxentius (Fig. 3.23) and in the Baths of Diocletian; and the latter had ribs. In the eleventh century both roofs were still as the Romans had left them. The cross vault of the Basilica of Maxentius collapsed in the fourteenth century, and in 1561 the tepidarium of the Baths of Diocletian (which had the cross vault) was converted into the Church of S. Maria degli Angeli by Michelangelo.

The third ingredient, the arcade, derived from the Palace of Diocletian at Spalatum (see Section 3.9). It was used consistently in Christian architecture, but the Gothic form became much lighter than the Romanesque.

The pointed arch had several significant advantages over the semicircular arch of Romanesque architecture. In the semicircular arch the rise was the radius, and the span, the diameter; that is, the span was always twice the rise. In the pointed arch this ratio could be varied at will over a wide range.

One of the greatest structural problems of the Gothic cathedrals was created by their enormous height and the use of arches and vaults that produced horizontal reactions. In this respect the Gothic arch had an advantage. Because its ratio of span to height was normally less than 2, it produced a smaller horizontal reaction.

The Gothic arch approximates the catenary arch in shape (see Section 5.3). The closer the arch or shell to the catenary, the lower the ratio of thickness to span

6.1

Igreja do Carma, Lisbon, built 1389–1423, and destroyed in the great earthquake of 1755. The slender pointed arches survived.

The Gothic Interlude

required (Section 6.6). The Gothic master builder was limited in his geometry (Section 6.4); therefore an arch that could not be set out with a compass would have been impractical. An arch composed of circular arcs is the nearest approximation to this most favorable form (Fig. 6.1).

6.2 THE STRUCTURAL FEATURES OF GOTHIC ARCHITECTURE

The overall dimensions of Gothic architecture were not remarkable. The span of the nave of the more famous cathedrals (although remarkable for their great height) is of the order of 14 m (45 ft) and rarely exceeds 16 m (51 ft). Gerona Cathedral has the largest span of any Gothic nave (22 m, or 73 ft, completed in 1458) because it has no aisles but was exceeded by the cross vault of the Basilica of Maxentius (25 m, or 83 ft) completed in A.D. 312. The dome of the Pantheon has a span of 44 m (138 ft), built in A.D. 123.

The tallest vaults are in France, both in the north: Amiens Cathedral, 43 m, or 140 ft, built in 1288, and Beauvais Cathedral, 48 m, or 158 ft, built in 1347. In England the tallest vault is in Westminster Abbey, 31 m, or 102 ft, built in 1260. By comparison, S. Sophia in Constantinople, completed in 537, has an interior height of 54 m (179 ft).

Even in overall height Gothic buildings did not set records. The spire of Salisbury Cathedral, the highest in England, built in 1258, has a height of 123 m (404 ft); the tallest still standing, the spire of Strasbourg Cathedral, built in 1439, rises to 142 m (466 ft). The tower of Beauvais Cathedral, which was built in 1569, had a height of 155 m (510 ft) but collapsed in 1573.

The Pyramid of Khufu, built about 2600 B.C., had an original height of 147 m (482 ft), and the Cross of S. Pietro in Rome, built in 1590, stands 138 m (452 ft) above ground.

In spite of these statistics, Gothic cathedrals convey a tremendous impression of height (S. Pietro looks big, Chartres looks tall) created by the upward surge of Gothic sculpture, an artistic rather than a structural achievement.

The greatness of Gothic construction does not lie in overall dimensions but in the refinement of structural detail. The Egyptian pyramids could not collapse. Roman buildings, in spite of their enormous size, were structurally overdesigned. Gothic cathedrals could, and did, fall. The arrangement of structural members, and their sizes, is therefore of great interest.

The overwhelming majority of Romanesque and Gothic cathedrals had pitched timber roofs. Small churches, but also some of greater importance, for example, S. Croce in Florence, had only one. In a cathedral church, however, a stone vault was built under the timber surface. There were two main purposes for this double-roofed structure.

Because the cathedral was the tallest building in any medieval city, it was likely to attract lightning, and for that reason alone the timber roof sooner or later caught fire. (Lightning conductors were invented by Benjamin Franklin in the eighteenth century.) Medieval fire-fighting methods were not so efficient as those of ancient Rome, and as the height of the churches increased interior protection became essential, for most cathedrals were built around a holy relic. The stone vault, in turn, needed protection from the weather. As the complexity of vaulting increased, water was no longer able to drain off the roof. In addition, it has been argued that a timber roof was needed to protect the interior from the weather during construction (which often continued for decades and even centuries) and for use in hoisting the stones of the vault into position (Ref. 5.6 and Fig. 6.5). Timber roofs tended to become steeper (Fig. 6.2) in the later Gothic cathedrals. This may have been a matter of style or the need for a more efficient hoisting frame for the more complex vaults of that period.

Timber roofs are subjected to lateral wind pressure, and the steeper the roof, the greater the pressure. This is true even if a horizontal tie is placed near the base of the

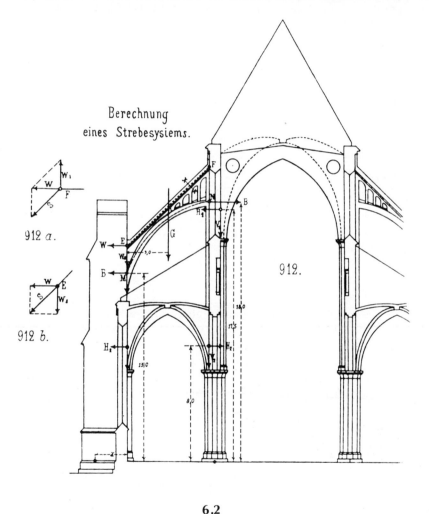

6.2

Forces acting at a typical cross section of a Gothic cathedral (Fig. 912, Plate 6, of Ref. 2, quoted in Ref. 5.13).

truss. The stone vault also produces a lateral thrust (see Section 6.5 and Fig. 6.9). In many Gothic cathedrals these two thrusts are handled separately, the first by an upper flying buttress and the second by a lower one (Fig. 6.2). These flying buttresses transmit the lateral thrust to a massive buttress on the far side of the aisle, within which the combined vertical and lateral thrusts are transmitted to the ground.

Most Romanesque and Gothic churches have a central division, called the nave, and two outer divisions, or aisles. The vault of the nave is supported on two rows of pillars that separate the nave from the aisles, and the vaults of the aisles are supported on one side by the pillars and on the other side by the outer walls, which are stiffened on the outside by buttresses (Fig. 6.2). As the Romanesque style changed to Gothic, the walls became thinner, and the windows in them much larger, so that the walls carried almost no load. The buttresses became much wider. In addition, the difference in height between the vault over the aisles and the vault over the nave became more pronounced. Thus it became necessary to build flying buttresses, and as the difference in the height between the nave and the aisles increased the flying buttresses had to be arranged in two layers. As we have already noted, the upper flying buttress was frequently placed in line with the thrust of the timber roof and the lower flying buttress in line with the thrust of the stone vault of the nave.

In the early Gothic cathedrals the vaults were quadripartite; that is, they were formed by two intersecting cylinders. Where the cylinders intersected there were ribs and where the ribs intersected there were bosses. The function of the ribs and bosses has been the subject of much controversy, but it is sufficient to note here that they provided a neat solution to the difficult geometry at the intersections and a means of setting out the vaults (Ref. 6.6). We shall consider the structural and constructional functions of the ribs in Sections 6.4 and 6.6.

In the later Gothic cathedrals the geometry of the vaults became more complex and the number of ribs increased. In the Perpendicular style, the latest period of English Gothic, the ribs developed into a fan consisting of an intricate network of small ribs; we shall consider the structure of fan vaults in Section 6.6.

6.3   THE SOURCES OF INFORMATION ON GOTHIC DESIGN

The general structural features of Gothic architecture are thus quite clear, but the manner in which the actual dimensions were determined is obscure and the sources of information are totally inadequate. This has produced a vast literature that is largely speculative and contradictory. We know far less about the design of the medieval cathedrals than about Roman or Renaissance architecture.

One possible reason was the low level of literacy, which in the twelfth century was still mainly confined to the clergy. King John, whose reign is contemporary with the rise of Gothic architecture, did not sign the Magna Carta because he could not write. Many of the Gothic master builders may have been able to read and write, but books were much scarcer than in Roman times.

The sketchbook of Villard de Honnecourt, about whom little is known otherwise, has survived and has been published in full with translations of the writing into English (Ref. 1.2) and German (Ref. 6.19). It consists of thirty-three sheets of parchment, with writing and drawings on both sides. In addition, there are annotations by

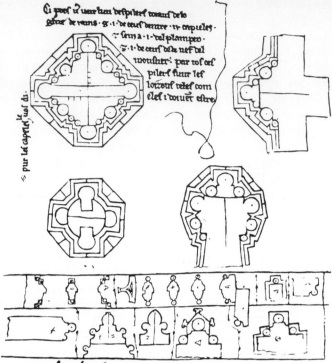

6.3

Sections of columns in Reims Cathedral drawn by Villard in the early thirteenth century (Ref. 1.2).

two other persons, called by Frankl (Ref. 6.3) Magister 2 and 3. Frankl devoted twenty pages to a discussion of the notebook and the annotations, and most writers on Gothic architecture make frequent reference to it. The notebook was probably written between 1225 and 1250, that is, the period of French High Gothic, and the annotations made by Villard's pupils.

Some of the drawings are skilful, in the manner of manuscript illustrations of the period and with an attempt at three-dimensionality. Some construction lines and centers of circles are still discernible. Several plans, sections, and elevations of Gothic cathedrals include those at Reims (Fig. 6.3) (which Villard, however, did not build). There are descriptions of some contrivances:

If you wish to make a hand warmer, you must first make a kind of brass apple with two fitting halves. Inside this brass apple there must be six brass rings, each with two pivots, and in the middle there must be a little brazier with two pivots. The pivots must be alternated in such a way that the brazier always remains upright, for each ring bears the pivots of the others. If you follow the instructions and the drawings, the coals will never drop out, no matter which way the brazier is turned. A bishop may freely use this device at High Mass; his hands will not get cold as long as the fire lasts. That is all there is to it. (Ref. 1.2, p. 66).

Another machine shows an eagle with internal pulleys, annotated "How to make the eagle face the Deacon while the Gospel is read" (Ref. 1.2, p. 129). None of the gadgets, however, equals in ingenuity those described by Hero of Alexandria in the second century B.C. (Section 2.3).

The hoisting machines and catapults illustrated are inferior to those described by Vitruvius (Ref. 2.3), but there is a saw that utilizes a rope, a pulley, and a weight for cutting off the tops of piles under water (Ref. 1.2, p. 131).

One diagram explains "how to cut the voussoirs of hanging arches"; voussoirs (blocks of stone) are cut at an angle so that they cannot fall out, as in a pendant fan vault (Fig. 6.17). Villard shows a wooden prop under the pendant with a rope for pulling it away (Ref. 1.2, p. 123). Another, annotated "how to set up two pillars without plumb line or level," shows an equilateral triangle. A page is devoted to fifteen small diagrams "extracted from geometry," but the geometry is very simple.

We have three treatises written between 1140 and 1144 (Ref. 6.3, pp. 3–24) by Abbot Suger of St. Denis (Section 6.1), widely regarded as the first Gothic cathedral. The abbot was the client and possibly the designer of this structure. Suger wanted to match the marble columns of the existing basilica-style church, built in the days of the Carolingian kingdom (see Section 5.1), and described how he considered obtaining them from the Baths of Diocletian, or other ancient baths in Rome, and moving them by sea to France and up the river Seine to Paris. Finally he found a suitable quarry, but there was the problem of handling the stone:

Whenever the columns were hauled from the bottom of the slope with knotted ropes, both our own people and the pious neighbours, nobles and common folk alike, would tie their arms, chests and shoulders to the ropes and, acting as draft animals, draw the columns up; and on the delivery in the middle of the town the diverse craftsmen laid aside the tools of their trade and came to meet them, offering their own strength against the difficulty of the road, doing homage as much as they could to God and the Holy Martyrs (Ref. 6.3, p. 7, reprinted by permission of Princeton University Press).

This voluntary help later became "the cult of the carts." Suger also described a storm that took place on January 19, 1143. Geoffrey, Bishop of Chartres, was at the time celebrating a special mass for the soul of King Dagobert. He was alarmed by the vibration of the arches and the roofing and "frequently extended his blessing hand in the direction of that part. . . . Thus the tempest, whilst it brought calamitous ruin in many places to buildings thought to be firm, was unable to damage those isolated and newly made arches, tottering in mid-air, because it was repulsed by the power of God." This account has been discussed by many Gothic historians (Ref. 6.3, pp. 13–17) and is open to different interpretations; however, none explains satisfactorily the construction or design of the vaults.

A tendency to interpret structural behavior in terms of divine intervention was one of the problems of the time. The central tower of Winchester Cathedral fell down in 1107, a few years after it was built. A chronicle gave as the reason that William Rufus was buried there seven years earlier (Ref. 5.3, p. 6).

At Durham in the twelfth century Bishop Pudsley built a Lady Chapel, but according to the contemporary chronicler Geoffrey de Coldingham

the walls having been erected to scarcely any height, the work at length yawned with fissures and gave manifest indications that it was not acceptable to God and his servant, Cuthbert. And having abandoned that work, he began another at the west end, into which women might lawfully enter, so that they who had not bodily access to the more secret things of the holy places might have some solace from the contemplation of them.

Because the walls were low, the cause of the trouble was probably a foundation failure (Ref. 5.3, p. 8).

The construction of the cathedrals, which, next to war, must have been the most expensive activity of the time, is mentioned in many chronicles, some of which, for example, by the monk Gervase on the rebuilding of Canterbury Cathedral after the fire of 1174, are quite detailed. A few contain pictures of men hoisting material or laying masonry (Ref. 6.1, pp. 25 and 30). None, however, gives any indication of the method of design.

Frankl discussed five later manuscripts by master builders (Ref. 6.3, pp. 145–154): An anonymous sketchbook in the Nationalbibliothek of Vienna, about 1450; a geometry by Heutsch, written probably in 1472; another by Schmuttermeyer, probably written in 1486; a text on the verticality of pinnacles by Roriczer in 1486; and an instruction book by Lacher, written in 1516. These explain setting-out procedures but not design.

Useful data can be drawn from surviving account books. A. J. Taylor (Ref. 6.26, pp. 104–133) has discussed the recruitment of labor in fourteenth-century Wales. Major building projects, like the medieval cathedrals and castles, required hundreds of skilled craftsmen who had to be brought from distant counties after the local supply had been exhausted.

In 1277 Edward I decided to build castles in Wales to aid him in his fight against Llewellyn. He hired an *ingeniator*, a designer of military installations, from Viennois in southern France, and sent two of his clerks to recruit labor through the local sheriffs. Altogether, 3000 people, including 800 carpenters and 300 masons, were brought from places as many as 180 miles (290 km) from the building site. Evidently the sheriffs were expected to supply the labor, but the men were paid normal wages. Taylor quotes a letter to the Bishop of Bath and Wells:

I have to inform you that 40 carpenters from the County of York came to us at Chester on the first Sunday after Trinity by command of our lord the King; also 150 diggers on the same day and from the same county. We began paying their wages from the Monday next ensuing, and have paid them nothing for the five previous days, i.e. the 5 days they spent in coming to Chester, in respect of which the Sheriff asks to be allowed the money he paid them for their expenses in accordance with the injunction to him (Ref. 6.26, pp. 115–116, by courtesy of the Hamlyn Group).

The men walked to Chester, with a mounted man as escort, but were allowed a cart to carry their tools. The lowest paid mason received 1 shilling 3 pence per week, and the master mason, 3 shillings per week; records also show bonus payments to the master.

The accounts give information on materials. Stone, lime, sand, and timber normally came from a nearby site, but if water transport was practicable distance was not so important. We have noted that Suger considered bringing columns from Rome for St. Denis and that stone from Caen in northern France was used for Canterbury Cathedral. In the accounts examined by Taylor the greatest distance for masonry and timber was 19 miles (30 km), but glass, lead, iron, and tin had to be brought from farther afield.

In the thirteenth century the master masons organized themselves in lodges or guilds, which were active in many cities, but the Strasbourg chapter acquired a special status and disputes were often referred to it.

The design of the cathedrals is still not completely understood. Not merely did the setting out of the plan and elevation require much skill, and we have some information on how this was done, but the sizing of the structural members was often critical (Sections 6.5, 6.6, and 6.7). Presumably a great deal of the knowledge was carried in the heads of the master masons. When a journeyman became a master, he was initiated into the mystery.

Medieval masons traveled great distances in their capacities as masters. European roads of their time were inferior and less safe than those of ancient Rome, the contemporary Inca empire in Peru, or China of the Yuan dynasty, but transport of materials was affected more than people.

Villard (Ref. 1.2) states that he went from France to Hungary in the early thirteenth century. William of Sens came from France to become master mason of Canterbury Cathedral in the twelfth century, and in the thirteenth Etienne of Bonneul went from Paris to Uppsala, Sweden, as master mason of the cathedral.

A succession of foreign experts was called to Milan to give advice on the cathedral, which, because of its great size (second only to Seville), gave frequent cause for concern. The cathedral was started in 1386; it is not known who designed it, but a model was made to which the builders largely adhered throughout the construction. In 1389 Nicholas de Bonaventura was called from Paris. In 1391 Master Giovanni was sent to Germany to show the drawings to the master mason of Cologne Cathedral. In 1394 Ulrich von Ensingen of Ulm, Germany, went to Milan, and in 1399 Jean Mignot (Giovanni Mignoto) arrived from Paris and remained as the cathedral's architect until dismissed in 1401. The cathedral had reached a critical stage, and several more experts were called in to advise during Mignot's term of office. The arguments were recorded in the documents of the time, published in Milan in 1877, and discussed by Frankl (Ref. 6.3, pp. 62–86).

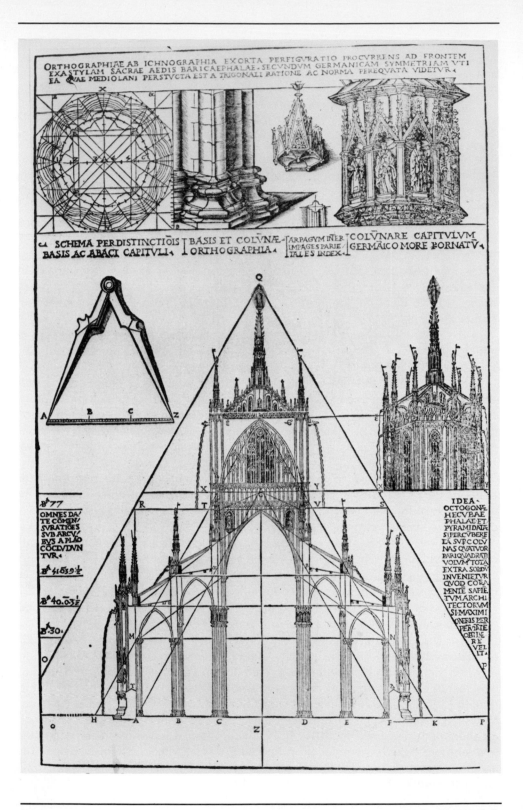

ORTHOGRAPHIAE AB ICHNOGRAPHIA EXORTA PERFIGVRATIO PROCVRRENS AD FRONTEM
EXASTYLAM SACRAE AEDIS BARICAEPHALAE·SECVNDVM GERMANICAM SYMMETRIAM VTI
EA· QVAE MEDIOLANI PERSTVCTA EST A TRIGONALI RATIONE AC NORMA PEREQVATA VIDETVR·

◁ SCHEMA PERDISTINCTIŌIS | BASIS ET COLVNAE· | [ARPAGVM INTER | COLVNARE CAPITVLVM
BASIS AC ABACI CAPITVLI· | ORTHOGRAPHIA· | IMPAGES PARIE | GERMĀICO MORE ▷ORNATV◁
 | | TALES INDEX·]

B̄·77
OMNES DA/
TE CŌMEN/
SVRATIŌES
SVB ARCV/
BVS A PLĀO
CŌCLVDVN
TVR·

B̄·41.ō39½

B̄·40.ō3½

B̄·30·

O

IDEA·
OCTOGONĀE
HECVBAE ET
PYRAMIDAĪE
SI PERCVBERE
EĀ SVB COLV
NAS QVATVOR
ŌARIQVADRĀTI
VOLVM̄ TOTĀ
EXTRA SOLIDV
INVENIETVR
QVOD CORĀ
MENTE SAPIĒ
TVM ARCHI
TECTORVM
SI MAXIMI
ONERIS PER
ŌPERITĀTE
OBIRE RE
VEL
IT·

Evidently these exchanges of views, rather than books that may have been lost, were the basis of design. Although the master masons talked to one another, they were not communicative in public, in which respect medieval architects differed significantly from those of Rome, the Renaissance, or today.

There is at least the possibility that the successful master masons got the right answer by applying rules that worked for reasons they themselves did not understand. Mignot repeatedly claimed that "ars sine scientia nihil est," by which he did not mean that art is nothing without science but rather that craftsmanship is useless without theory (see Section 1.2). The theoretical dispute was between those who wanted the elevation of the cathedral designed *ad quadratum* by means of a series of horizontal and vertical lines contained within a square and those who wanted it designed *ad triangulam* by means of a series of triangles (see also Section 6.4). Evidently the inclined lines in Fig. 6.4 conveyed an impression of the lines of thrust and may well have been helpful in buttress design. In addition, they served to fix the height of the cathedral. The minutes of May 1, 1392, contain the question: "Whether the church itself, that is not counting the tower over the crossing, is to be raised according to the square or according to the triangle?" The answer: "They declared that the church itself be raised up to the triangle or to the triangular figure and not above it." Frankl interpreted *figura triangularis* as the equilateral triangle (an angle of $60°$) in Fig. 6.4 and *triangulum* as a Pythagorean triangle with sides 3, 4, and 5 (an angle of $53°7'$) (Ref. 6.3, p. 69). *Ad quadratum* implied an equal width and height or a triangle with an angle of $63°28'$ at the base (Mainstone, Ref. 6.4).

Both Frankl and Simson (Ref. 6.5) discussed the question of proportions, the latter in great detail. It is quite possible that harmonic proportions were used in the design of Gothic cathedrals, particularly those of the later period, for the work of St. Augustine and Boethius (Section 2.2) on music was well known in the Middle Ages. No definite evidence supports these suppositions, however.

It seems likely that designs were originally made on the flat. The cost of drawings is frequently mentioned in accounts, and some quite detailed ones on parchment survive. An elevation of Strasbourg Cathedral is preserved in the Historical Museum in Berne, Switzerland, and the completed design closely conforms to it. Elevations still in existence for the cathedrals in Cologne, Orvieto, Siena, and Vienna are also reported. There are fifteenth-century drawings for the spires of Ulm in Germany and Burgos in Spain, which were not built at the time but were used in the nineteenth

6.4

Elevation of Milan Cathedral in Cesariano's edition of Vitruvius, published in Milan in 1521. In that year the nave vault was completed and the octagonal cupola was under construction. The spire was built in 1756–1779. The main facade was erected in 1805–1813, and the last of the spires was completed in 1858. The cathedral had been started in 1386. Cesariano thus showed an incomplete design, and moreover some of the dimensions of the cathedral as built are not correctly drawn. The main triangles are equilateral, as are those in the sketches of the fourteenth century (Ref. 6.3).

century to complete the cathedrals (Ref. 6.1, pp. 36, 37, and 42). Parchment was a valuable material. The drawings for a cathedral would have needed the skins of a sizable flock of sheep, and it is therefore likely that they were cleaned and the material reused as soon as the work was completed. Full-size details were probably set out on a tracing floor (Section 6.4).

Models were mentioned frequently, and it is probable that the making of a large model was normal practice at an early stage in the design. None survives, however. Most would have been made of timber, but Antonio di Vincenzo spoke of a model of brick and plaster for S. Petronio in Bologna, built in 1390 and lost in a fire in 1423. This model was so large that one could walk inside. Frankl (Ref. 6.3, p. 299) has estimated its length as 18.9 m (62 ft).

The existing evidence is discussed in detail because it leaves one important question unanswered: how did the master masons determine the structural sizes, particularly the ratio of the thickness to the span of the vault (Section 6.6)? There is no evidence to support the contention that the Gothic master builders had some secret knowledge of statics which has been lost. The knowledge of mechanics at their disposal, at any rate during the early Gothic period, was less than that of ancient Greece when the Parthenon was built, ancient Rome when the Pantheon was built, or of Byzantium when S. Sophia was built. It would be surprising if they had not formulated some empirical rules based on the collapses during construction which appear to have been common and the less frequent collapses after completion which naturally caused more concern. These rules do not appear in any document described in the literature on Gothic architecture. The surviving manuscripts have been examined so thoroughly in the nineteenth and twentieth centuries that it is unlikely that any rules dating from the thirteenth, fourteenth, or fifteenth century will be found.*

It may be that the rules, if they existed, were regarded as *ars* (craftsmanship) and not as *scientia* (theory) and therefore unworthy of a written record, or it may be that the masters preferred not to commit them to parchment or even mention them in discussions of design such as those recorded in Milan.

The records show that geometry was the basis of design for both plan and elevation, and the latter may have offered some guidance for shaping the vaults and aligning the buttresses (see Fig. 6.4).

The design of the cathedrals was done by skilled specialists, as it is today, although an interested client may sometimes have offered suggestions or imposed his views, as he does today. The designer was generally the master mason as well. Master masons traveled widely and design concepts were conveyed by personal contact rather than in published works.

Some unskilled work was done by voluntary labor motivated by a desire to perform good deeds that would count on the Day of Judgment or perhaps by sheer

---

* The only exception is a rule attributed to Rodrigo Gil de Hontañon, a Spanish architect who died in 1577, which is claimed to give the depth of the ribs as a fraction of the span and a specific number for the ratio of distance between pillars and the depth of the ribs varying from 30 to 20. This was mentioned by Mainstone (Ref. 6.23, p. 306) and Frankl (Ref. 6.3, p. 536). Both saw it in an article by Kubler, published in the *Gazette des Beaux-Arts*, 1944, p. 135. Kubler got his information from a Spanish book, published in 1868, and its author saw it in another book which reproduced a Spanish text of 1685. Gil de Hontañon would himself be a post-Gothic architect. The rule, if correctly reported, is post-Gothic, and could be an original rule or a subsequent rationalization.

enthusiasm for a great work of art; there may also have been some forced labor, but most workmen, skilled and unskilled, were recruited in the same way as in the nineteenth century and paid normal wages. Materials were obtained nearby or transported by water.

6.4   THE CONSTRUCTION OF THE GOTHIC CATHEDRALS

The setting out was done by geometry, not by means of a graduated scale, for two reasons. The making of accurate subdivisions on a scale required technology that the Middle Ages did not possess and measures had not been standardized. It is likely that each city kept a standard foot or yardstick against which craftsmen could, or were obliged to, check their own measuring rods, but it probably varied from place to place.

Medieval masons and carpenters had compasses, shown in many contemporary illustrations, with which a measure could be transformed, multiplied, and divided with considerably accuracy. Once the base line had been set out in local feet, everything else was derived from it by means of geometric constructions. Roman geometry was probably never entirely lost, and the full text of the *Elements of Euclid* was translated into Latin in the early twelfth century (see Section 2.2). Books had to be copied laboriously by hand, but by the mid-thirteenth century masons probably had acquired a considerable knowledge of practical geometry.

The vertical was set out with a plumb bob, as it is today. Horizontal alignments were taken with a straightedge fixed at right angles to a plumb bob, which is as accurate and almost as convenient as the modern spirit level.

The setting out was sometimes done on parchment but more commonly on a plaster floor, called a trasura. The cost of this floor appears in several accounts (e.g., Ref. 6.26, p. 164) and some can still be recognized from the markings (e.g., at Wells Cathedral; Ref. 6.1, p. 33, and at York Minster, Ref. 6.29). From these tracings templates were made or the block was cut directly on them.

The stone blocks themselves had marks on them to locate them within the structure. Sir George Gilbert Scott, a noted nineteenth-century restorer of Gothic cathedrals, was generally able to locate them:

They made the upper surface of the boss-stone horizontal, to serve as a sort of drawing-board on which to draw the plan of the intersecting ribs. I have tested this in several instances. In the western part of the nave in Westminster, there being no outer thickness of stone vaulting, the boss-stones appear, and their surfaces are horizontal. On sweeping away the accumulation of dust and rubbish which covers them, I found, sure enough, the centre and side lines of all the ribs carefully drawn upon them.

In the lierne vaulting at Ely, though there has been an outer thickness of stonework, it was cleared away in the last century for the sake of lightness, so that the boss-stones, once concealed, are now visible. On clearing them from obstructions, I again found, as at Westminster, the lines of the ribs, here much more complex, carefully set out upon the top of the stones. Each of these little stone tables, in fact, has drawn out upon it the bit of the full-size plan of the vaulting which its surface would contain (Ref. 6.6).

In Gothic cathedrals the largest blocks of stone were small compared with those of ancient Egypt, Greece, or Inca Peru. Judging by the contemporary illustrations made by Villard and other writers (see Section 6.3), hoisting equipment was much more limited than in ancient Rome. There is no evidence that compound lifting tackles of the type described by Hero of Alexandria and Vitruvius were ever used (Sections 2.3 and 2.4), nor was it possible in the manner of ancient Egypt and Greece to build a ramp of earth and drag or roll the stones up that ramp; the cathedrals were too high. Consequently the stones had to be small. As Ruskin correctly observed, this is a more sophisticated structural form than that of ancient Greece (Section 3.2).

Medieval mortar was weak and slow setting (Ref. 5.6, Appendix G, pp. 262–265) and contributed little to strength during construction. The medieval concrete with which buttresses and piers were generally filled (see Section 5.7) was similarly much weaker than Roman concrete.

Scaffolding presented another problem. Although timber was still plentiful, the length of the trees and their weight imposed limitations. It is unlikely that the interior of a Gothic cathedral was filled with a forest of scaffolding, as one might be tempted to do today.

Viollet-le-Duc (Ref. 6.7) has argued that the walls, piers, and ribs were built with flying scaffolds, that is, scaffolds fixed into holes left in the masonry. This would have been a simple method of supporting the working platforms, but the holes would have had to be left or subsequently filled in when no longer required.

It may be reasonably assumed that Gothic scaffolding was flimsy and by present European or American standards not particularly safe. Flimsy scaffolding is still used in many developing countries; for example, in South East Asia from Indonesia all the way to India.

Fitchen has suggested that the timber-roof structure was built before the stone vault. Thus it protected the masonry during construction (which might take many years) from rain and wind and provided a hoisting frame for the stones of the vault (Fig. 6.5). Windlasses used for hoisting survive in the timber roofs of Durham, Peterborough, and Norwich Cathedrals (Ref. 5.9, pp. 67–73). The increasing steepness of Gothic timber roofs may have been a matter of style or it may have been due to the need to provide a stronger hoisting frame. In the case of St. Stefan's in Vienna, built from 1304 to 1450, the rise of the timber roof is actually greater than the height of the nave vault (the present roof is a replica of the one destroyed in 1945).

Most authorities on Gothic architecture (Ref. 5.6) have reached the conclusion that Gothic vaults were not built on full timber formwork, as done in Roman concrete vaults and domes and as would be done in a modern vaulted structure. Several arguments support this view. A vernacular tradition still exists for the building of masonry vaults and arches (Fig. 6.6) in the Mediterranean countries and Latin America. They are generally erected with only sufficient scaffolding to support the weight, never with close boarding; it is only because of the primitive formwork that these methods are still economically viable in the twentieth century.

It seems likely that Gothic vaults are part of a long tradition of building vaults freehand, a view supported by the minor irregularities of Gothic webs which would be less likely to occur with close boarding, although some could result from deformation after construction.

Fitchen (Ref. 5.6) described two methods by which this could be done (Fig. 6.7). Two masons, back to back, would add the courses from each side, meeting in the middle. Only the joints between the stones needed to be supported by planks on end,

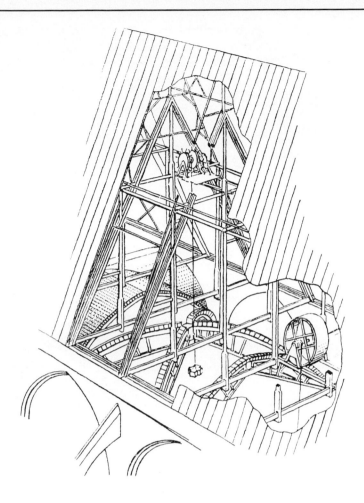

6.5

Method of handling stone voussoirs suggested by Fitchen (Ref. 6.8, p. 286). The roof shown is that over the nave at Reims Cathedral and the hoisting wheels are shown supported on the transverse tie beams of the timber roof.

and they, in turn, rested on the centering for the ribs which was still in position (Fig. 6.7*a*). Alternately, the newly laid stones would be kept in position by a weighted rope attached to the timber roof (Fig. 6.7*b*). This possibility was first suggested by de Lassaux, architect to the King of Prussia in the early nineteenth century (Ref. 6.9).

There has been much discussion of the functions of the ribs in Gothic construction (see also Sections 6.6 and 6.7). It has been argued that they have a sculptural quality and should be regarded as part of the artistic concept, that they exist to prevent the buckling of the thin stone webs, that they form a structural frame rather like the iron frame of the Crystal Palace of the 1851 Exhibition in London and support the web of

6.6a

Roof in the Jewish Quarter of the Old City of Jerusalem.

the stone vault, that they are necessary to provide a neat junction between the component parts of the vault, and that they are required for the construction of the vaults, given the technical facilities available to medieval builders.

All these arguments may be correct, but the last seems the most important. It is likely that the ribs were constructed first on centering only sufficient to support their own weight and that the stone webs were then added while the rib centering was still in position. Thus the ribs were required to locate the webs but need not necessarily have supported them. It is further likely that the webs, but not the ribs, were built freehand. It does not, of course, follow that the same methods were used in all cathedrals.

6.6*b*

Modern repairs to a twelfth century mosque in Kayseri (Turkey). Support is provided only at the joints (see also Fig. 6.7a).

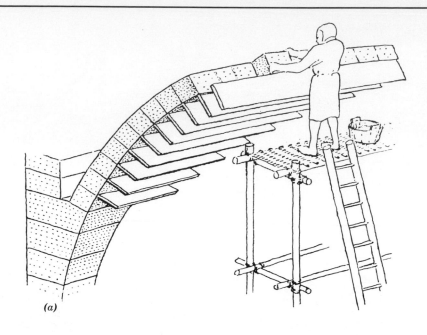

(a)

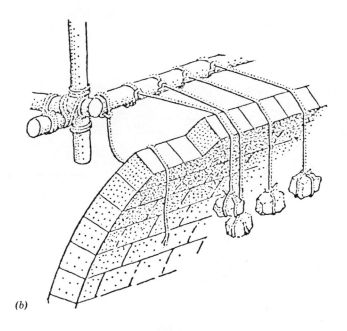

(b)

6.7

Possible methods of building the shells of the Gothic vaults, considered by Fitchen (Ref. 5.6, pp. 140 and 182).

The Gothic Interlude

(a) The mason stands on a light platform, and lays the stones from below on planks-on-edge supporting them at the joints. The method of supporting these planks on the centering for the ribs is not shown (see also Fig. 6.6b).

(b) In the method suggested by Lassaux, the newly laid stones are held in position by weighted ropes tied to a framework which is fixed to the roof structure.

6.5   BUTTRESSES AND PINNACLES

One would naturally wish to start a discussion of Gothic structure with contemporary views on the subject. Unfortunately none has survived.

The significance of the buttresses and pinnacles was appreciated by Sir Christopher Wren, who repaired several Gothic cathedrals. He used buttresses in the design of St. Paul's (see Section 7.4).

In the nineteenth century the Gothic Revival produced a new interest in the structure of cathedrals. By that time the theory of structural mechanics had been developed to the extent that any statically determinate structure could be analyzed; therefore it was applied to Gothic architecture (Refs. 6.2, 6.10, and 6.11). The elastic concept of masonry design was probably formulated by Navier in 1826 in his *Résumé* (see Section 8.3). The first exposition in English appeared in a book by Moseley (Ref. 6.12), published in 1843. The concept that the line of thrust must lie within the middle third is implicit in this analysis, but it was stated explicitly by Rankine (Ref. 6.13). The rule is illustrated in Fig. 6.8.

The elastic theory of the nineteenth century made two assumptions about the behavior of the buttresses: (1) that the mortar joints have no tensile strength; this is entirely reasonable in view of the low strength of medieval mortar and is also made in all subsequent theories; and (2) that for structural safety the joints must not open under load; this is unduly conservative and accounts for the heavy look of Neo-Gothic masonry designed in accordance with this theory.

We have to contend with an inclined thrust from the roof, and we assume that the flying buttress has been correctly aligned to transmit this thrust (we examine this assumption further in Section 6.6). We must now transfer the horizontal component of this thrust to the ground, and we do this by adding the weight of the pinnacle. Using the parallelogram of forces (which we discuss in Section 7.4), we obtain a new resultant which has a smaller angle with the vertical. We now add the weight of the upper section of the buttress, and the resultant comes closer to the vertical. By adding more weight and widening the buttress successively, we eventually bring the thrust to the ground (Fig. 6.9). As long as the resultant thrust remains within the middle third no tension develops and the joints do not open.

Nevertheless, if we analyze the medieval Gothic cathedrals, we find that many have pinnacles and buttresses that are much too small, according to this theory, but have not failed. There have been reports of failures, but the dimensions of the members that proved inadequate are generally lacking.

6.8

Model illustrating the middle-third rule. When a load is placed outside the middle third, tension develops on the opposite side and the joint opens up (*Architectural Science Laboratory, University of Sydney*).

The analysis of buttresses by the middle-third rule assumes that the material has adequate strength, and even in small and heavily loaded buttresses the compressive stresses are at most one-tenth of the compressive strength of the material (Ref. 6.11). At the top of a buttress, however, the shear strength of the stone could be critical. The failure occurs diagonally, as shown in Fig. 6.10, if there is no pinnacle. On the other hand, a compressive stress produced by even a small pinnacle would be sufficient to prevent a failure. A pinnacle can improve structural safety even when its weight is insufficient to change the direction of the resultant thrust significantly.

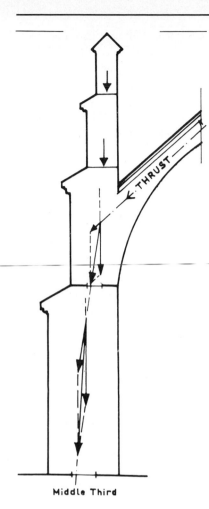

Middle Third

6.9

Analysis of buttress by the middle-third rule. At any section the resultant of the thrust and weight of masonry above it must lie within the middle third if opening of the joints in the masonry is to be avoided.

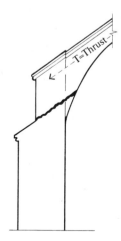

6.10

Failure of buttress due to the absence of a pinnacle. If the thickness of the buttress is insufficient, the masonry fails by shear (diagonal tension). This can be prevented by adding a pinnacle of sufficient weight to eliminate the tensile stress.

There is a great deal of literature, reviewed by Frankl (Ref. 6.3) and Heyman (Ref. 6.11), on the purpose of flying buttresses and pinnacles. Viollet-le-Duc (Ref. 6.7) has argued that both were structural features; Gothic cathedrals were unstable without buttresses and their form was purely determined by functional considerations. On the contrary, Abraham (Ref. 6.14) has claimed "that neither flying buttresses nor pinnacles were necessary. Many a French cathedral had none, and acquired them only when restored by Viollet-le-Duc." Without an elaborate program of experimental research neither claim can be substantiated.*

It is interesting to note a report on Chartres Cathedral, written in 1316 and quoted by Frankl (Ref. 6.3, p. 59, reprinted by permission of Princeton University Press):

"Further we have examined the flying buttresses which support the vaults; the joints are sadly in need of sealing and repair, and if no one will do it at once, there may be great damage." This suggests that the person who wrote these lines in the fourteenth century believed that the flying buttresses supported the vault of Chartres Cathedral.

Viollet-le-Duc, who lived from 1814 to 1879, was a little younger than Paxton, designer of the Crystal Palace, and a little older than Jenney, who designed the first steel-framed building in Chicago (Ref. 1.3, Section 2.6). It was natural therefore that he should interpret Gothic architecture in terms of the iron skeleton frame to which it does indeed bear a much greater resemblance than any other architecture before the nineteenth century. He produced in his Lectures on Architecture (Ref. 6.17) numerous designs that featured iron struts and iron roofs in otherwise medieval designs of enormous size (Fig. 6.11). He was an effective propagandist for functional design and Gothic architecture, and unlike romantic Neo-Gothicists like Pugin and Ruskin he could see no conflict between the two. As Inspector of Historic Monuments he re-

---

* An experimental program is entirely feasible, but expensive, and probably not justifiable as pure research. Two sets of model investigations of Gothic cathedrals have been reported. The first, by Mark, was undertaken at Princeton University. Cross sections of the Chartres and Bourges Cathedrals (Ref. 6.15a) and of Beauvais Cathedral (Ref. 6.15c), including their system of buttresses and flying buttresses, were made from epoxy plastic to scales of 1:180 and 1:144, and tested photoelastically; it was observed that the pinnacles suppressed dangerous tensile stresses at the junction of the buttress with the upper flying buttress in the elastic range. Mark also tested a three-dimensional photoelastic model of the nave vault of Cologne Cathedral (Ref. 6.15b) to a scale of 1:50.

The second, by Fumagalli (Ref. 6.16, pp. 77–79 and 94–96), was undertaken to deal with the problems of Milan Cathedral (see Section 6.3), but the concern was with the stability of the tiburium, i.e., the octagonal drum supported on sixteen pillars, which in turn carries the spire. The problem was created by the water requirements of the city which led to the dewatering of underlying waterbearing strata and resulted in uneven settlement.

The investigation was carried out on an elastic model made to a scale of 1:15 from epoxy resin filled with sand and polystyrene grains. The second model of a pillar only was made from the same materials as the cathedral pillars, namely a core of Serizzo granite with a cladding of Candoglia marble, to a scale of 1:4.7. The quality of the mortar and the method of laying was as close as possible to that used by the medieval masons. This test led to two interesting conclusions: (1) the marble, having a modulus of elasticity 2.4 times that of the granite, took more than its share of the load and failed relatively early by vertical fissuring and separation from the granite core; (2) using the comparatively rough methods of medieval masons, the pillars had only one-third the strength expected from masonry pillars built by modern methods with perfectly finished surfaces.

Neither investigation settled the question whether flying buttresses are structurally required, but evidently this could be established beyond reasonable doubt if one were prepared to go to the trouble of an ultimate-strength model test to a scale of, say, 1:5.

The Gothic Interlude

6.11

"Vaulting of large spaces by iron and masonry." Design suggested by Viollet-le-Duc in the middle of the nineteenth century (Ref. 6.17).

stored many French Gothic cathedrals, and added flying buttresses in some cases where none had existed before. To that extent Abraham is correct.

On the other hand, Gothic cathedrals have had partial collapses from the moment they were built until the nineteenth century. There was still a major accident in 1861 when the tower and spire of Chichester Cathedral telescoped into the church. After restorers such as Viollet-le-Duc and Scott (Ref. 6.6) had done their work there were no further collapses.

Because the Gothic master masons had no means of structural sizing, except possibly empirical rules (see Sections 6.3 and 6.7) and tended toward a light design (perhaps for economic, perhaps for artistic or religious reasons), there were bound to be some grossly underdesigned structural members that collapsed immediately (we do not know their dimensions, but we do know that there were numerous failures); there were some also that had only a small factor of safety and might fall down later if circumstances were unfavorable (e.g., a high wind, decay of stone of poor durability, a minor repair that altered the load distribution, or settlement of the foundations that caused a collapse at a later time); there were some that had a factor of safety of the kind we now consider appropriate; and there were some that were grossly overdesigned.

In an overdesigned structure or even one with an adequate factor of safety the vault would not necessarily fail if the flying buttresses were missing or removed; conversely, the flying buttresses would not necessarily fall down if the vault were removed. Thus we have Gothic ruins that according to Viollet-le-Duc, lack supporting members but are still standing. Gothic architecture was just not that accurate.

It is hard today to determine how the Gothic cathedrals were designed, but it takes no advanced knowledge of mechanics to understand the flying buttress. As the cathedral vaults got higher and the walls thinner, it was to be expected that the walls would deform outward, and any sensible builder would endeavor to prop them up temporarily with an inclined piece of timber. A flying buttress is merely a permanent masonry structure of the same kind.

The case for the structural function of the pinnacle is not so strong. The upward surge of Gothic sculpture could by itself account for the pinnacle as an artistic rather than a structural feature. On the other hand, some of the Early English cathedrals, such as Lichfield in Staffordshire (Fig. 6.14), have formidable pinnacles in conjunction with an appreciable difference in the height of the nave and aisles which make a pinnacle necessary. In late Gothic, particularly in the Perpendicular style, pinnacles became stylized and structurally useless, but this paralleled a similar degeneration in the function of the ribs and was the last stage of Gothic structure on the decline.*

As we noted in Section 6.2, flying buttresses in the larger cathedrals were in two layers, frequently parallel and at an angle of a little more than 36°, corresponding to a right-angled triangle with sides 3, 4, and 5 (the Pythagorean triangle). This was easily set out by geometry, but it is not particularly efficient as a means of transmitting the thrust of the vault to the ground. In Ely Cathedral, north of Cambridge, the flying buttresses in the choir were set out parallel in the mid-thirteenth century according to the Pythagorean triangle (Fig. 6.12). In 1322 the central tower, built in the twelfth

---

* The terms "degeneration" and "decline" are intended to apply purely to the structure. Some of the most beautiful Gothic buildings belong to this period; for example, Henry VII Chapel in Westminster Abbey and the Chapel of King's College, Cambridge.

The Gothic Interlude

6.12

Section through the choir of Ely Cathedral. The original buttresses of the midthirteenth century are shown on the right-hand side. The new buttresses built in the fourteenth century are shown on the left. The earlier design is geometrically simpler, but the greater structural efficiency of the later design is evident (Ref. 5.3).

century, collapsed and brought down the flying buttresses on the north side. They were rebuilt in the fourteenth century at a much steeper angle, the upper even steeper than the lower. This was an unusual design for Gothic flying buttresses but clearly more efficient. One does not really need an advanced knowledge of mechanics to determine the approximate alignment, but the sizing of the buttresses was difficult, as we shall see presently. At Ely the master mason took no chances during the reconstruction and the buttresses were clearly oversized.

An unusual feature of the collapse of the Ely tower was its replacement by an immense timber vault and lantern, built by William Hurley between 1328 and 1340. It was a masterpiece of medieval carpentry (Ref. 5.9, pp. 82–84) and much safer than a stone vault, but, of course, not fireproof. Evidently the authorities at Ely did not want to risk a further collapse.

Another interesting set of buttresses was built at Wells Cathedral in the west of England in the late twelfth and early thirteenth centuries. In the early fourteenth the height of the central tower was increased. The pillars at the crossing of any Gothic cathedral were subjected to a certain amount of bending because the lateral thrusts from the vaults of the crossing and the naves were imposed on them at different levels and were not self-balancing as in a normal arcade. When the extra weight of the tower was added, the pillars at Wells started to bow inward, and in 1338, about a century and a half after they were built, the inverted arches (Fig. 6.13) were added as buttresses. This remarkably skillful design enhanced the interior space and blended perfectly with the old construction. It also illustrates the slow pace at which medieval architecture developed.

Our discussion so far has been based on the proposition that joints in the masonry must not open up. If we design a buttress on that basis and then increase the inclined thrust, the joints do open but the buttress will not collapse so long as the line of thrust lies wholly within the buttress. The limit is reached when the line of thrust touches the outline of the buttress (see Section 7.6). Thus it is possible to support a thrust above that calculated by the design illustrated in Fig. 6.9, provided the resultant force everywhere remains within the section. When the resultant falls outside the middle third, but within the section, some joints will open up but the structure will not collapse. Because the Gothic cathedrals were built slowly, this opening up of the joints was a process spread over years and the joints could be filled by repointing. It is impossible to tell now which joints of Gothic buttresses may have opened slightly since their original construction. Lime mortar is soluble, and there can be few joints today that have not been repointed at some time.

There are two methods for designing the buttresses in this state. One is to allow the resultant to fall outside the middle third but inside the "middle half"; this is an arbitrary rule. A safer design method is to determine the thrust that would cause collapse and then make sure that the actual thrust is less. In modern reinforced concrete design two conditions are specified: the structure must be able to resist one and one-half times the dead load (i.e., the weight of the building material); and as an alternative condition it must be able to carry one and one-quarter times the dead load plus one and one-half times the wind load. This seems a reasonable approach to the design of a Gothic buttress. As we shall see, the load factor is not so important as in modern design because we are not concerned with stresses in the stone but with the stability of the buttress; this depends only on the line of thrust.

6.13

Arched buttresses between the nave and the crossing in Wells Cathedral, erected in 1338, one and a half centuries after the pillars had been built. There are similar buttresses on the other three sides of the crossing.

We have so far assumed that the flying buttress is weightless and correctly aligned to transmit the thrust of the vault or roof structure. The first is obviously not correct, and the second can be satisfied only by a designer who can calculate the inclination of the thrust from the roof structure, which the Gothic master builders were probably unable to do. The flying buttress therefore was not a simple strut but subject to bending.

Flying buttresses are often so slender that one's first reaction is one of surprise that they remain supported at an apparently precarious angle. When analyzed, many fail to conform to the middle-third criterion. All that have remained standing, however, have a line of thrust wholly within the flying buttress.

The analysis can be simplified by using the safe theorem of limit design. Heyman (Ref. 6.27) stated this as follows:

If a line of thrust can be found which is in equilibrium with the external loads and which lies wholly within the masonry, the structure is safe.

Heyman added that the line of thrust found in order to satisfy this theorem need not be the actual one; any line in equilibrium with the external loads and lying within the masonry is sufficient to ensure stability.

In the same paper Heyman postulated a further theorem:

If on striking the centering for a flying buttress, that buttress stands for 5 minutes, then it will stand for 500 years.

This statement assumes that the loads are static and is therefore not applicable to wind loads, thermal movements, and so on (see comment in Section 6.7).

Heyman's method explains the structural behavior of the thinner flying buttresses which has previously puzzled structural engineers. His analysis of Lichfield Cathedral, previously mentioned, is shown in Fig. 6.14.

The two collapses of Beauvais Cathedral, in the north of France, have been the subject of much speculation and both have by some writers been attributed to inadequate buttressing. Several publications have dealt with these collapses and have been reviewed by Heyman (Ref. 5.13). The building of Beauvais Cathedral began in 1247 and the apse and choir were finished in 1272. In 1284 the vault collapsed.

A part cross section, as drawn by Viollet-le-Duc, is shown in Fig. 6.15. It will be noted that an intermediate buttress reduces the slenderness ratio of the flying but-

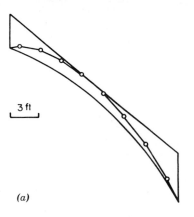

3 ft

(a)

6.14

Lichfield Cathedral in Staffordshire

(a)   Line of thrust for the flying buttresses of Lichfield Cathedral, as
      determined by Heyman (Ref. 6.27).

(b)   Cross section according to Banister Fletcher (Ref. 3.20, Fig. 4.19F).
      There is only one layer of flying buttresses at an angle of approxi-
      mately 41°. The pinnacles are substantial and are correctly aligned
      with the inside face of the buttress where a diagonal tension failure of
      the masonry would commence.

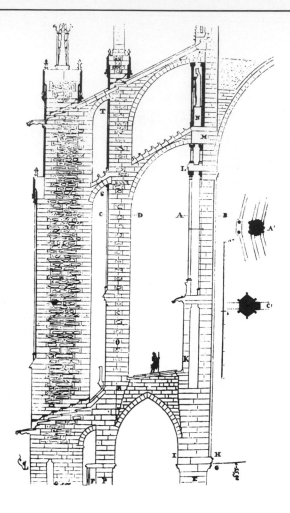

6.15

Part cross section of Beauvais Cathedral according to Viollet-le-Duc (Ref. 6.7, article "Construction").

tresses and that an additional flying buttress projects from the intermediate to the outer buttress at the level of the aisle vault.

Viollet-le-Duc thought that the buttresses were adequate and that the failure was initiated by excessive shrinkage of the mortar in the pier marked B in Fig. 6.15. The shrinkage of medieval mortar, remarked on by a number of historians (Ref. 5.6, Appendix G), would have had two consequences: it would have loosened the block M which carried a heavy statue N, otherwise freestanding, and it would have transferred additional load to the two slender columns A, which would have fractured as a result. The lintel L then would have broken and the block M would have slid out. This block would have transferred the thrust of the vault to the lower flying buttress and the vault would then have collapsed.

The Gothic Interlude

Heyman quoted several other explanations and analyzed the flying buttresses he found adequate. He then concluded that the collapse began with some trivial accident. Mark (Ref. 6.15c) has offered a different explanation, based on creep in the mortar.

No contemporary accounts of the failure or reasons for the collapse remain, but when the roof was rebuilt the length of the bays was halved; that is, it was reduced from 9 m (30 ft.) to 4.5 m (15 ft.), thus doubling the number of buttresses. This suggests that the master mason or the Cathedral Chapter thought that the buttressing had been insufficient.

An interesting sidelight on Beauvais is a midnineteenth century comment (Ref. 6.24) by Professor Richard Brown, an architect:

At this time the pillars being placed too far apart, the vaulted roof threatened to fall in, which actually took place after means had been adopted to support it by iron braces and chains, to hold the side walls together.

Iron ties were used in a number of cathedrals but apparently as a corrective device rather than as part of the original design. The practice was also common in earlier Muslim architecture.

The choir had been rebuilt by 1337, but for a century and a half work was interrupted by the Hundred Years' War between France and England. The English occupied Beauvais for part of that time. Work resumed in 1500, and by 1548 the transept had been completed. It was then decided to construct the tower over the crossing before the nave had been built. There was some question whether it should be a wooden or masonry tower, and the decision was in favor of stone. The tower was built from 1564 to 1569 and collapsed in 1573.

Beauvais was tall in all its dimensions. The height of the choir (there was no nave) was 48 m (158 ft), compared with 47 m for the nave of Amiens Cathedral, the next highest in France, and 31 m for that of Westminster Abbey, the tallest in England. The height of the tower was 155 m (510 ft), compared with 142 m for Strasbourg, the tallest still standing, and 123 m for Salisbury Cathedral, the tallest spire in England. It was therefore well beyond the normal practical experience of the Gothic master masons. Furthermore, the tower lacked the lateral support that the nave would have provided.

Heyman (Ref. 5.13) stated that the tower gave cause for concern almost immediately. Two of the king's masons reported in 1571 that the four main piers at the crossing were all out of plumb. Those on the choir side were four French inches out of plumb, and those on the other side (where the nave should have been) were out of plumb by 6 and 11 in. The masons thought 2 in. was acceptable, for the piers were well constructed. They recommended immediate erection of two nave bays, strengthening of the pier foundations, and construction of temporary walls between the piers. For two years the Chapter sought further advice and on April 17, 1573, agreed to put the work in hand. The tower collapsed thirteen days later. It was not rebuilt and the nave was never added.

The failure could have been due to foundation settlement, but it is quite likely that the masonry was cracked by the enormous tower (see Ref. 6.16, pp. 94–96). Gothic masonry was not so strong as modern masonry because the interior surfaces were not so well finished, because it usually contained a weak core of rubble, because the mortar lacked strength, and because there was appreciable shrinkage contraction in the mortar joints. Moreover, it would have been difficult to ensure that the piers were loaded concentrically, for recent experiments with concrete columns have shown that even small eccentricities of loading can produce great increases in stress. The wind pressure on the tower would have added to the bending stresses.

A number of reconstructions of Beauvais Cathedral show how it appeared before the collapse of the tower (e.g., Ref. 5.13). The tower looks too big for a cathedral without a nave, and it seems to be unsound in design artistically as well as structurally.

An interesting sidelight on the collapse of the spire of Beauvais Cathedral is the report on the state of Salisbury Cathedral made by Sir Christopher Wren in 1669:

As this addition of a spire is a second thought, the artist is more excusable for having omitted buttresses to the tower; and his ingenuity is commendable for supplying this defect by bracing the wall together with many large bands of iron within and without, keyed together with much industry and exactness; and besides those that appear, I have reason to believe that there are diverse other braces concealed within the thickness of the walls; and these are so essential to the standing of the work that, if they were dissolved, the spire would spread open the walls of the tower, nor could it stand one minute. (S. Wren, *Parentalia*, pp. 303–306, quoted by Hamilton, Ref. 3.25.)

6.6 GROINED VAULTS AND FAN VAULTS

Heyman has given the equations of the membrane theory for a thin-groined vault of semicircular shape; that is, with a rise a and a span $2a$ (Ref. 6.27). This vault, used in reinforced concrete at the air terminal in St. Louis, is sufficiently close to the pointed groined vault of Gothic architecture for determination of the reactions.

The reactions calculated by Heyman are shown in Fig. 6.16. In a hemispherical dome buttresses are not needed because the hoop tension absorbs the horizontal reaction (see Section 3.9). In cylindrical and groined vaults horizontal reactions are unavoidable, and in modern reinforced concrete structures they are usually absorbed by ties (see Ref. 1.3, Section 6.3). This was occasionally done with iron in Gothic architecture (see Section 6.5). There were three arguments against the use of iron: (1) it was still expensive; (2) it might corrode even if covered with lead and thus burst the masonry; and (3) the horizontal ties would interfere aesthetically with the verticality of the Gothic vaults.

The horizontal reactions are $2\omega a^2$, where ω is the weight per unit area of shell. They must therefore be absorbed by buttresses. It should be noted that the horizontal reaction is required not at the base of the vault but $0.466a$ above it. For a cylindrical vault with a 15-m (45-ft) span this is 3.3 m (10 ft). In a pointed vault the reaction would be slightly less and slightly displaced. Heyman points out that flying buttresses at St. Denis, Beauvais, and Amiens were all a little too low, but those of Notre Dame in Paris and of Lichfield and Reims were correctly placed.

At right angles the horizontal reactions were balanced by the adjacent shells except at the end of the nave, where buttresses were again required. During construction the horizontal reaction to be supplied by the adjacent shell not yet built must

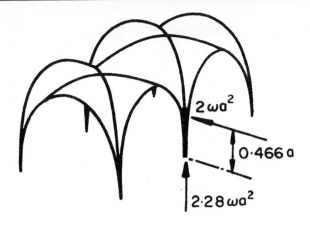

6.16

Heyman's analysis of the reactions required by a cylindrical groined vault, formed by the intersection of two semi-circular vaults of diameter 2 a. In a pointed vault the horizontal reactions are slightly less, and their point of action is slightly lower (Ref. 6.27).

also be absorbed by a temporary structure. Fitchen (Ref. 5.6) reports that hooks have been found in some Gothic cathedrals (e.g., Westminster Abbey) which might have been needed for fixing iron bars or chains to provide a temporary tie during construction. Alternatively, timber ties could have been used.

Heyman then calculated the forces at the groins. In all shells high stresses occur wherever there is a sharp change in curvature and the groined vault is no exception. Heyman found that the force reached a maximum at an angle of about 55° from the crown, that is, roughly two-thirds along the groin, and that the force was 1.3 ωa^2. This is quite substantial, and the shell must therefore be made sufficiently thick or reinforced at the groins. Thus ribs are structurally required.

To investigate the shell itself we draw the line of thrust (see Section 7.6). If it lies wholly within the thickness of the vault, the shell is just stable. For a typical pointed groined vault Heyman calculated the thickness required as a/50. For a nave width of 14 m (45 ft) this gives a thickness of 150 mm (6 in).

Fitchen (Ref. 5.6, Appendix E, pp. 256–259) has listed vault thicknesses, which can, of course, only be estimated except on rare occasions when a vault is cut for repair or by war damage. The figures he gave for groined vaults range mostly from 350 mm (14 in.) to 200 mm (8 in.), except that Viollet-le-Duc claimed thicknesses as low as 100 mm (4 in.). The sixteenth-century fan vault of King's College Chapel is reported to have a thickness ranging from 2 to 6 in. (50 to 150 mm), but that is a different problem, discussed below.

It was not uncommon to use a lightweight stone, such as chalk (at Salisbury) or tufa (in the Rhineland) for the high vaults; these stones have relatively low strength, but

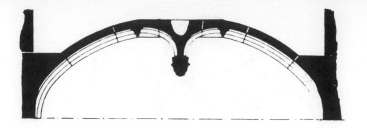

6.17

A pendant can be held in position by sloping the joints outwards, so that stone from which the pendant is cut acts like a wedge (Ref. 6.18).

because the compressive forces in the masonry, except at the groins and springings, were quite low no problems were presented.

At the junction between vault and pillar the stresses would increase in a vault of uniform thickness; therefore it was necessary to increase the thickness. In many Gothic vaults the conoid-shaped section above the pillar was filled with concrete.

At the other extreme of Gothic vault construction is the fan vault in which the curvature changes smoothly and the ribs are purely decorative (Ref. 6.20). Fan vaults sometimes had a pendant (Fig. 6.17) which was held in position as a wedge by sloping the joints outward and was a deadweight on the vault. Figure 6.18 shows a

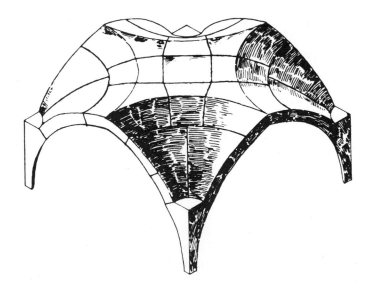

6.18

Idealized shape of fan vault (Ref. 6.18).

fan vault for a square column grid. The central portion is a spandrel only slightly curved and supported on four conoids. The lower part of the conoids was filled with concrete in most Gothic fan vaults. Heyman (Ref. 6.28) has given the solution for conoid membrane shells. The stresses are quite low, but the hoop forces become tensile near the top of the conoids if they carry merely their own weight; tensile forces are not admissible in an unreinforced masonry shell.

If, however, the shell is given an additional vertical load, the hoop forces become compressive throughout. The spandrel provides this weight, and a heavy pendant in the spandrel may even be helpful for the stability of the shell. Heyman drew the following conclusion:

Effectively, then, a late fifteenth-century designer specializing in fan vaults (King's College Chapel was vaulted in 1512–15 by John Wastell) could pick any likely-looking profile, could decorate the surface of the vault with non-structural ribs, could insert on the centre line of the nave a series of heavy bosses or pendants and could pay relatively scant attention to the external buttressing system, all with complete assurance that his structure would be stable (see comment in Section 6.7).

The point of application of the horizontal reaction is not as critical in the fan vault as it is in the ribbed groin vault; there is an appreciable horizontal reaction, however, which in King's College Chapel was resisted directly by large external main buttresses. Because the chapel consisted only of a nave, without aisles, flying buttresses were not required.

6.7 A PERSONAL EVALUATION

On a subject as controversial as Gothic structure it is perhaps appropriate for an author to state his own views explicitly, even though they are probably discernible under the objective veneer of the previous sections.

Gropius, echoing Nietzsche, distinguished between Apollonian and Dionysian, between serene and passionate architecture. Gothic tends toward the latter category and its sculptural exuberance can be compared only with that of the Hindu architecture of India. In the Baroque architecture the Dionysian aspect is in the layout and decoration; it hardly affects the structure. In Gothic architecture the structure is a part of the sculpture and is remarkable for its daring rather than its rationality. It therefore must be evaluated as a form of art rather than an engineered construction.

Gothic cathedrals have a special quality that Neo-Gothic architects have, on the whole, not been able to recapture. Not everybody likes Gothic, but few people have failed to be moved on first entering one of the great Gothic cathedrals. Neo-Gothic cathedrals rarely provide the same emotional experience. Some are splendid, most are nicely finished, and all are perfectly safe; but something is missing.

In the eighteenth century, as we shall see later, the theory of structural mechanics had reached the stage at which substantial parts of Gothic architecture could be analyzed, and this was done in the nineteenth. If one insists that the resultant must fall in the middle third to prevent the joints from opening up, then many Gothic cathedrals are unsafe; yet a great many that have failed to pass this test are in sound condition and have been so for several centuries.

I doubt, however, if a responsible building authority, acting within the philosophy of current building codes, could in a modern building accept tension in unreinforced masonry. At present we admit limit design for steel structures but not for reinforced concrete (except for the special and limited case of the yield line theory). Could we then allow it for plain concrete or unreinforced masonry?

I am greatly indebted to those engineers who have applied limit analysis to Gothic architecture, particularly Heyman, because they have demonstrated how a structure actually behaves. However, I believe that Heyman is too sanguine about the factor of safety required for unreinforced masonry structures and about the effect of settlement, temperature movement, and creep. Considering only those collapses known with certainty, it is evident that many structures fell which according to limit analysis were perfectly safe. The failures ceased only after the restorations of the nineteenth and twentieth centuries. It may well be that the increase in structural sizes which resulted from some of these restorations ruined the artistic concept of some of the cathedrals and could have been avoided by the use of hidden steel reinforcement or prestressing.

Various writers on Gothic architecture have referred to mathematicians who were consulted about its design, and a few instances have been documented. It seems certain that their evidence was used for the overall geometry, not for structural sizing. No mathematician or physicist could have given any useful advice on structural design before the fifteenth century, and it is doubtful that it would have been helpful before the seventeenth.

Several authors have mentioned the possibility that models were used for structural design. There is a record of at least one model, for S. Petronio in Bologna, which was of masonry and would have been large enough to yield useful information if it had been accurately loaded and tested to destruction. This was not done however; the model was lost in a fire. The other models appear to have been smaller and made of wood. Looking at a structure three-dimensionally is always helpful, but it is doubtful that these models could have been used for experimental analysis. Their purpose was to help the designer with the overall geometry, to explain the design to the client, and to preserve continuity, for the construction often continued for decades and sometimes for centuries.

I am convinced that rules were used to limit the ratio of depth to span, or to specify minimum thickness, as they were in building codes in the nineteenth and the early twentieth centuries and still are on domestic construction. They could have been arrived at empirically by observing that certain dimensions were safe, whereas others were associated with collapse (or with movement when the centering was loosened, thus warning of a possible collapse). Even today builders take the striking of the centering for large arches and shells very seriously, carefully observing deformations as it is removed and replacing it if there is visible movement.

Why do we not know what these rules* were? It may be that they were not regarded as worthy of a written record. There is no medieval book comparable to Vitruvius. No medieval author has told us how stone should be selected or mortar proportioned. Perhaps these practical points were regarded as *ars,* or craftsmanship, which was not to be put in the same class as *scientia,* the geometry of the overall design. Similarly, we have no Greek texts on the practical aspects of building.

* See footnote in Section 6.4.

Another possible explanation is that the master mason was sworn to secrecy by his lodge.* He may have been given these rules when he became a master and was not permitted to divulge them except to a new master at a special ceremony. The ceremonial of freemasonry may have a practical antecedent. Modern consulting engineers do not publish all their design details when they write a paper on one of their projects, but if a similar building is required they will be pleased to design it.

Master masons were brought from great distances at a time when communications were poor, and consultants were frequently needed when cathedrals were in trouble. In fact, the Gothic era seems to have made more use of foreign experts than the Renaissance. Evidently the master masons had some special knowledge that could be shared only in personal consultation. That vaults and flying buttresses became notably thinner between the twelfth and the fourteenth centuries suggests that the designers had learned from experience and from one another.

Although Roman and Renaissance hemispherical domes were self-balancing, Byzantine domes and Gothic vaults required buttressing. If the vault was high, the pillars of the nave would have had to be enormous without flying buttresses across the aisles, particularly if there was a steep timber roof or tower to transmit wind loads. In many cases flying buttresses were essential. Similarly, some weight, such as a pinnacle, was required to prevent a diagonal tension failure of the junction of the flying buttress with the outer buttress. I agree with Viollet-le-Duc on both points and disagree with Abraham.

Having discovered the need for flying buttresses and pinnacles, Gothic designers used them as part of the sculpture. They became stylized and in some late Gothic churches, superfluous.

Ribs were required in the groin vaults to provide reinforcement at the sudden change in curvature and were probably essential for constructional purposes. I consider that Abraham is correct in claiming that they did not form a framework to support the webs, as suggested by Viollet-le-Duc. Ribs do not support the webs and webs do not support the ribs; they do, however, reinforce one another.

In late Gothic architecture the ribs became more numerous and changes in curvature less abrupt. The structural necessity was gradually transformed into a decoration. In fan vaulting there is a continuous change of curvature without sharp corners and ribs could be omitted.

* Rosenberg (Ref. 6.21) states that the Bishop of Utrecht in Holland was killed by a master mason in 1099, because the bishop had induced the mason's son to divulge the method of laying out the foundations of a church.

CHAPTER SEVEN

The

Renaissance

It seems to me that those sciences which are not born of experience—that is to say, those sciences whose origin or process or end does not pass through the five senses—are vain and full of errors.

LEONARDO da VINCI

The Renaissance produced a great change in architectural design but only a gradual development in the use of materials. There were, however, dramatic structural innovations which we shall consider with particular reference to the three great domes: S. Maria del Fiore in Florence, S. Pietro in Rome, and St. Paul's in London.

We shall also examine the prehistory of structural mechanics during the Renaissance and the development of proportional rules.

7.1   THE RENAISSANCE OF GREEK SCIENCE AND ROMAN ARCHITECTURE

The Renaissance (i.e., rebirth) was a forcefully proclaimed revival of interest in classical science, art, and architecture, which began in Florence in the early fifteenth century,* and the consequent rejection of the "barbarism" of Gothic. Renaissance architecture was not, in fact, a copy of Roman architecture. It is possible to date the start of this *movement*, first in Florence, then in other northern Italian cities, then in other parts of Europe, with some precision. The renewal of *interest* in Greek science (see Section 5.4) and Roman architecture before the Renaissance was a gradual process; following the Renaissance there was a considerable overlap with Gothic architecture, even in the same city and sometimes in the same building.

The rejection of Gothic occurred in the writings of many Renaissance architects, including those of Brunelleschi, but it was perhaps nowhere put more clearly than in Filarete's treatise, written in Milan between 1461 and 1464:

I too was once pleased by modern buildings, but as soon as I began to enjoy the antique ones I grew to despise the modern. In the beginning, if I had anything built, I usually chose the modern manner, because my lord my father had used that fashion.

His princely patron then asks:

Why do you think this science has declined so that antique usage has been discontinued, even though it was beautiful?

I will tell you, my lord, it happened for this reason. Architecture declined as letters declined in Italy; that is, spoken and written Latin became more gross until fifty or sixty years ago, when minds became more subtle and were reawakened to the past. As I say, it was a gross thing. The same happened to this art through the ruin of Italy brought on by the wars of the barbarians who desolated and subjugated it many times. Then, too, many customs and rites came from the other side of the Alps. Because no great buildings were built, since Italy had become poor, men were no longer experienced in these things. As men lost experience, their knowledge became less subtle. Thus the knowledge of these things was lost. Then when anyone wanted to build any building in Italy, he had recourse to those who wanted to do the work . . . .

These modes and customs they have received, as I said from across the mountains, from the Germans and the French. For this reason ancient usage was lost (Ref. 7.11, pp. 175–6).

The reason round arches are more beautiful than pointed. It cannot be doubted that anything which impedes the sight in any way is not as beautiful as that which leads the eye and does not restrain it. Such is the round arch. As you have noticed, your eye is not arrested in the least when you look at a half-circle arch. . . .

* The Italian term *quattrocento* is frequently used in the English-language literature; it means the four hundreds (of the second millennium).

The pointed is not so, for the eye, or sight, pauses a little at the pointed part and does not run along it as it does on the half circle. This is because it departs from its perfection. . . .

You could perhaps say that pointed arches are strong and satisfactory. This is true, but if you make a round arch, that is, a half circle, with a good haunch, it too will be strong. To prove that this is true: I have seen large round arches in Rome that remained strong, especially in the baths, in the Antoniana, and many other buildings. If the Romans had doubted their strength at all, they would have made two arches one above the other, but they would never have used any of these pointed arches. Since they did not use them, we should not use them (p. 103).

The rejection of the *arte moderna* of the barbarians and the revival of the *arte antica* of ancient Rome was seen by Filarete as an aesthetic decision, although partly motivated by nationalism. We also find in his treatise an emphatic statement that *scientia* (theory) is more important than *ars* (craftsmanship), which we have already noted in Section 6.3.

The new rational approach to building, the freer discussion of theory, and the return to Roman methods of craftsmanship promoted a rapid advance in building construction; however, the abandonment of the pointed arch for purely historical or doctrinaire reasons created structural problems that were solved brilliantly with entirely different devices by Brunelleschi and Wren (see Sections 7.2 and 7.3).

Filippo Brunelleschi, who was born in 1377 and died in 1446, is credited by some critics, for example, Clark (Ref. 7.1), with having invented the Renaissance single-handed by designing the Foundling Hospital (*Spedale degli Innocenti*) in Florence, begun in 1421, and the dome of Florence Cathedral, built between 1420 and 1434; the latter is discussed in the next section.

Gothic construction, however, continued into the sixteenth century. The tower of Beauvais Cathedral (Section 6.5) was started only in 1564, and in Milan, a mere 250 km (150 miles) from Florence, the building of the Gothic cathedral continued throughout the age of the Renaissance (see Fig. 6.4).

The ground has been prepared for the Renaissance by the revival of learning in the twelfth and thirteenth centuries and the recovery of a number of Greek books lost to the West in the Dark Ages. The full text of Euclid's *Elements* (Section 5.4) became available in Latin in the early twelfth century, and the universities established themselves in the thirteenth and fourteenth.

In the early fifteenth century a number of scholars, by consulting different libraries, endeavored to reproduce the precise texts of Latin classics. Gian Francisco Poggio, a papal secretary, produced a definitive version of Vitruvius in 1415, based mainly on the since-lost manuscript in the library of St. Gall, a Swiss monastery and a major medieval seat of learning. It was only a day's ride from Constance, where Poggio attended the Council of 1414. We do not know whether *Vitruvius's Ten Books* was important in ancient Rome, but in the later Middle Ages it had been accepted wherever a copy was available. From the fifteenth to the seventeenth centuries its authority was almost without challenge. As late as 1715 Colin Campbell chose *Vitruvius Britannicus* for the title of his own book.

The invention of printing completely transformed the supply of books. In the Middle Ages a book was sometimes literally worth its weight in gold. Except for the Bible, books were scarce in Europe, and the handwritten texts frequently inaccurate.

The technique of printing may have filtered through from China, where movable type had been in use since the eleventh century, or it may have been the independent invention of Johann Gutenberg, who started experimenting with printing presses in 1438 and in 1456 produced the first printed bible. Vitruvius was among the earliest secular writers to appear in print, probably in 1487. By the midsixteenth century copies of his work were readily obtainable in northern Italy.

The invention of gunpowder in the fourteenth century may also have derived from China, although the European mixture was much more powerful. It ended the dominance of the armored knight and the medieval castle and brought feudalism to an end. The cities became the centers of power and wealth, and commerce and manufacturing assumed a new significance; for example, Brunelleschi's client in the construction of Florence Cathedral was the Opera del Duomo, a council financed mainly by the Woollen Guild.

Gunpowder made it necessary to invent a new type of fortification to withstand the ever-increasing power of artillery; Leonardo da Vinci was only one of many who turned their minds to this important problem, and the result was a great improvement in the quality of construction.

The voyages of discovery, made possible by more accurate astronomical charts and trigonometric tables obtained from the Muslims (Section 5.4) and the arrival in Europe of the magnetic compass, also started in the early fifteenth century. The property of magnetism was known in ancient Rome and is mentioned by Lucretius in *De rerum natura*; indeed, the name derives from the town of Magnesia in Thessaly, where magnetite was found. Its use in navigation is a Chinese or Arab invention brought to Europe by returning Crusaders in the thirteenth century. Prince Henry the Navigator, a son of King John I of Portugal, made his first voyage in 1419, and Nuna Tristão reached Guinea in 1446. The fall of Constantinople to the Turks in 1453 accelerated the search for a new route to the Indies; Christopher Columbus discovered the West Indies in 1492 in a Spanish ship, and Vasco da Gama reached the real India in a Portuguese ship in 1497. The new colonies provided a vast increase in the wealth of Europe, which affected both the quantity and quality of building. The center of power gradually shifted to the countries on the Atlantic coast, but the Renaissance buildings of the fifteenth and the early sixteenth centuries are all in Italy.

7.2 THE DUOMO OF FLORENCE

The use of the word dome, meaning cupola, dates from 1656 (Ref. 7.2). The word derives from *domus dei*, the house of God; thus the German *Dom* and the Italian *duomo* signify a cathedral. The Dom of Cologne has spires and the Duomo of Milan is also Gothic, but the most famous is the Duomo of S. Maria del Fiore in Florence; its cupola is not merely the first great structure of the Renaissance but its most outstanding technical accomplishment as well.

It is not known who originated the concept of placing a huge octagonal cupola on the Duomo. It may have been part of the original concept of Arnolfo di Cambio, who started to build the nave and the aisles in 1296, to replace an old cathedral dating from the seventh century.

The foundations for the great octagon were built by Neri di Fioravante in 1366. In the following year a large brick model built alongside the cathedral was approved and views of an octagonal dome, one with and one without a drum, appear in frescoes in other churches of Florence (Ref. 7.12, p. 109). It would seem that the general shape of the dome was established before Filippo Brunelleschi's time, but whether Neri had a scheme for erecting a cupola over the great span of 42 m (138 ft) is not known.

In the light of modern experience, we would regard the roofing of this great octagon without steel or reinforced concrete as the major issue to be resolved. Indeed, few consulting engineers of the present time would feel competent to undertake the design. The discussion in fourteenth-century Florence, however, revolved around two more immediate problems. The first was the question of cost; Brunelleschi was appointed designer of the dome largely because he was able to suggest a method of construction that required little scaffolding and therefore saved a great deal of money.

The second problem was the question of style. The Duomo had been started in the Romanesque style, but the span and the height of the nave were such that it was difficult to avoid the use of Gothic buttresses. The problems have been discussed in several contemporary documents (Refs. 7.13 to 7.15). The issue was at least partly considered in political terms. Florence was at that time controlled by the guilds and belonged to the Guelph faction, which tended to support the Pope and Italian nationalism. Milan, ruled by the Visconti family, belonged to the Ghibellines who supported the Holy Roman (but, in fact, German) Emperor; there were many Germans at the Milanese court and the style of its Duomo was Gothic (see Section 6.3). The antipathy between Milan and Florence therefore encouraged continuation in the Romanesque style. Stylistic consistency was probably a lesser issue, for many churches begun in Romanesque finished in Gothic without aesthetic conflict.

The span of the nave of the Duomo is roughly the same as that of S. Croce, another of the great Romanesque churches of Florence. In S. Croce painted king-post timber trusses were used to support the roof without stone vaults. (These trusses are not fireproof, but they do not impose a horizontal thrust on the supporting walls.) In the Duomo it was decided to use a cross vault, but without Gothic buttresses, when the construction of the nave vault was begun in 1366. Instead, Giovanni Ghini installed iron ties, which are clearly visible in the interior, to absorb the horizontal reaction.

Because of the reluctance of some Italian designers to use Gothic buttresses, iron ties were not uncommon in Italy and were also used in Saracen architecture, with which Italian architects were familiar. In addition, there was the ancient Roman tradition for the use of iron clamps (see Section 3.9).

The main objection to iron was the danger of splitting the stone if the iron rusted and consequently expanded. Exposed iron ties are, moreover, unsightly and would certainly have detracted from the appearance of the cupola.

The *tamburo*, or drum, which supported the dome, was started in 1410; it is possible, but not certain, that it was Brunelleschi's design. In the Byzantine manner the drum is surrounded by semidomes on all sides of the octagon except those bordering on the nave, but they are too small to absorb the entire horizontal reaction of the cupola. It is possible that horizontal reinforcing rods were inserted in the masonry above the semidomes (Ref. 7.14, p. 20), a view supported by the absence of vertical cracking in those parts.

In 1418 the *Opera del Duomo*, the council set up by the guilds to control the construction, announced a competition "for a model or design for the vaulting of the

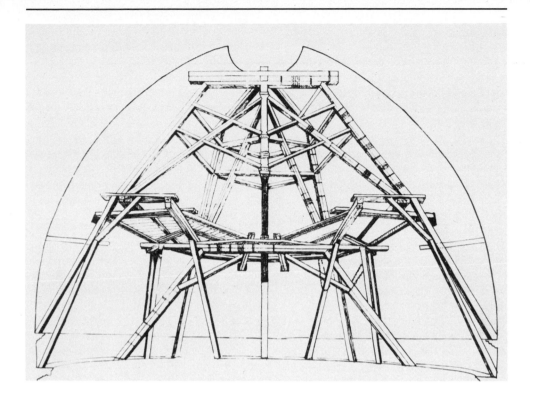

7.1

Scaffold, supported on the drum, designed by Brunelleschi for the construction of the cupola. From an account by G. B. Nelli in 1755.

main Cupola . . . for a scaffold or other thing or any lifting device pertaining to the construction. . ." (Ref. 7.14, p. 27). Brunelleschi submitted a specification. By the end of 1418 four masons had constructed a model to a scale of approximately 1:12 which showed his design of the cupola and the method of erecting it without complete formwork, as would be done at the present time (Fig. 7.1). The actual construction of the masonry was to be freehand, presumably as suggested by Fitchen for Gothic construction (Fig. 6.7a).

Brunelleschi's structural design included two revolutionary features. One was the division of the cupola into inner and outer shells, justified on the grounds of preventing the ingress of water that might damage frescoes on the inside of the dome; he may also have considered the extra stiffness that a double dome would provide. This device has been followed in many subsequent long-span structures to resist bending moments in the shell.

The second feature was the inclusion of a series of "chains" of timber and iron and stone and iron (Fig. 7.2). These were described in the specification submitted by Brunelleschi to the Opera del Duomo in 1420 (Ref. 7.14, pp. 32–41, reproduced by permission of M.I.T. Press; my conversion of measurements):

First, the inner cupola is vaulted in five-part form in the corners. Its thickness at the bottom point from which it springs in 3¾ braccia (2.1 m or 7 ft). It tapers so that the end portion surrounding the upper oculus is only 2½ braccia (1.5 m or 5 ft).

A second, outer cupola is placed over this one to preserve it from the weather and to vault it in more magnificent and swelling form. It is 1¼ braccia (0.7 m or 2½ ft) thick at the bottom point from which it springs, and tapers to the upper oculus, where it is only ⅔ braccia (0.4 m or 1¼ ft) thick.

The empty space between the two cupolas measures 2 braccia (1.2 m or 4 ft) at the bottom. This space contains the stairs to give access to all parts between the two cupolas. The space terminates at the upper oculus, 2⅓ braccia (1.4 m or 4⅔ ft) wide.

There are 24 ribs (*sproni*), eight in the corners and sixteen in the sides. Each corner rib has a thickness of 7 braccia (4.2 m or 14 ft) at the outside. Between the corners there are two ribs in each side, each 4 braccia (2.4 m or 8 ft) thick at the bottom. The ribs tie the two vaults together. They converge proportionally to the top where the oculus is.

The said twenty-four ribs, with the said cupolas, are girdled by six circles (*cerchi*) of strong sandstone blocks. These blocks are long, and are well-linked by tin-plated iron.* Above said blocks are chain rods of iron (*cantene di ferro*), all around said vaults and their ribs. At the start solid masonry has to be laid, 5¼ braccia (3.2 m or 10½ ft) high, then the rib outlines must be followed separately.

The first and second circles are 2 braccia (1.2 m or 4 ft) high, the third and fourth circles are 1⅓ braccia (0.8 m or 2⅔ ft), the fifth and sixth 1 braccia (0.6 m or 2 ft) high, but the first circle, on the bottom, is also reinforced with long sandstone blocks laid transversely, so that the inner and the outer cupolas rest on said blocks.

At the height of every 12 braccia (7 m or 23 ft) or thereabout, of said vaults, there will be small arches (*volticciuole a botti*) from one corner rib to the next intermediate rib, going around said cupolas. Below said system of small arches from one rib to the other are big oak beams (*catene di quercia*),† which tie the said ribs. Above each of said timbers is a chain of iron (*catena di ferro*).

The ribs are entirely built of grey and tan sandstone, and the covers or faces of the cupolas are entirely of tan sandstone, tied to the ribs, up to the height of 24 braccia (14 m or 46 ft). From there upwards the masonry will consist of brick or porous stone, as may be decided by the man who has to build it then. At any rate, it will be a material lighter than rock.

The rest of the specification dealt with nonstructural matters. In fact, brick was used for the upper part of the dome, and Parsons (Ref. 7.15, p. 592) commented that the bricks are of uniformly good quality, well burned and with straight edges. Bricks that were both well burned and undistorted by firing were rare in the fifteenth century, but Brunelleschi personally supervised their manufacture. The bricks varied in size from 300 × 150 mm (12 × 6 in.) upward and all were 50 mm (2 in.) thick. Parsons found the mortar exceedingly hard and thought that a hydraulic cement of the type used in ancient Rome had been employed. The bricks were laid in a herringbone bond which made it possible to dispense with supporting centering (Ref. 7.12, p. 113).

* Prager and Scaglia note that a gray or light blue sandstone quarried north of Florence was used. The six chains were increased to seven, and lead-lined iron was substituted for tin-plated iron.

† Actually, chestnut was used, 337 mm (13¼ in.) deep × 305 mm (12 in.) wide, interconnected with rectangular plates of oak, two at each joint and each 120 mm (4¾ in.) thick, fastened together by one large and six smaller iron bolts.

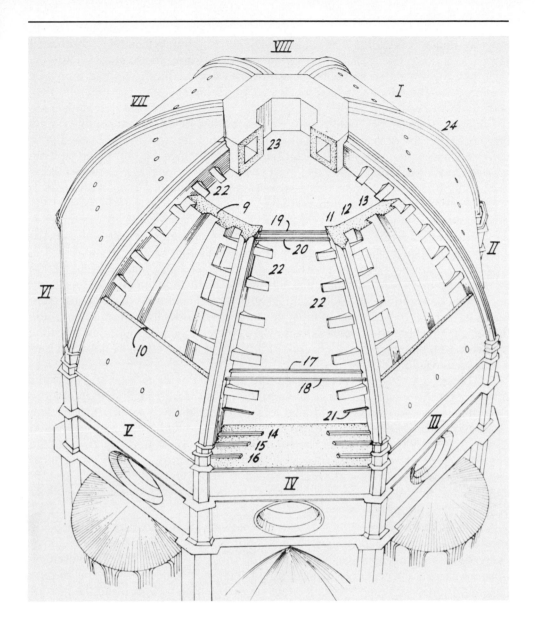

7.2

Masonry fabric of cupola of the Duomo of Florence, drawn by G. Rich in 1969 (Ref. 7.14, p. 35; reproduced by permission of M.I.T. Press). I-VIII: eight sides of the dome, partly removed to show interior framework. 9: inner masonry shell. 10: outer masonry shell. 11: main ribs. 12, 13: intermediate ribs. 14, 15, 16: stone chains at lower level. 17, 18: stone chains at intermediate level. 19, 20: stone chains at upper level. 21: timber chain. 22: horizontal arches. 23: oculus on top of dome. 24: ridges over main ribs.

As we noted in Section 3.9, the stresses in a dome along vertical (i.e., meridional) lines are all compressive. Parsons drew the line of thrust for the main ribs (Fig. 7.3), including the weight of the lantern, and this lies just within the middle third (see Section 6.5). The maximum stress occurs at the springings (base) of the main ribs, calculated by Parsons as 51,400 psf (357 psi or 2.46 MPa), modest for sandstone.

Evidently Brunelleschi could have created a more favorable stress condition if he had converted the octagon into a circle by using pendentives in the Byzantine or Saracen manner (see Section 5.2). He accepted the challenge, however, and built the greatest octagonal cupola of all time. In a hemispherical cupola (see Section 3.9) the horizontal (hoop) stresses are compressive from the crown to a hoop at an angle of about 52° with the crown and then change to tensile (Fig. 3.30). If the base of the dome were able to expand freely, the maximum hoop force would occur there; in the Duomo of Florence this was prevented, and the maximum tensile stress occurs above the springings. Parsons considered that Brunelleschi's timber chain (Fig. 7.2) was strategically placed to resist the bursting force which he calculated as 2,229,000 lbf (9.92 MN). He also decided that the timber chain was much too small, but apparently he was unaware of the stone chains. Including the five lower ones, there might have been enough iron to resist this force. We know the construction of the timber chain, which was exposed in the space between the two shells, but the stone chains were built into the masonry and the amount of iron used in the links and rods is not known (see Ref. 7.14, Fig. 10).

The other critical section occurs where the angle of the joints (see Fig. 7.3) is just sufficient for the weight to overcome the friction and shear strength of the mortar in the joint; that is, the stone would tend to slide under its own weight. Brunellischi suggested in the final paragraph of his specification (omitted in the passage quoted above) that the dome be built to a height of 30 braccia (17.52 m or 58.4 ft), and the position then reviewed. Parsons considered that this corresponded to the critical section—above it full formwork would be needed—but Mainstone (Ref. 7.12, p. 113) thought that the use of a herringbone bond in the brickwork made it possible to dispense with supporting centering.

Ample evidence shows that Brunelleschi not only understood qualitatively the behavior of this huge and completely novel dome but was able to obtain quantitatively correct data. Unfortunately we do not know how he did it. The surviving documents contain accounts of the cost of the work and details of the dispute between Brunelleschi and his coarchitect Lorenzo Ghiberti (who made the famous doors of the Baptistry) but no explanation of the structural design. It is most improbable that Brunelleschi arrived at the solutions theoretically. Even Leonardo da Vinci, who was born seventy-five years later, would have been unable to solve this problem (see Section 7.4). The dome of the Pantheon in Rome (Figs. 3.20 and 3.29) could not have served as a guide. Although it is nearly the same size, it relies on mass, whereas the Duomo has a sophisticated frame with tension reinforcement. It is known that Brunelleschi made a careful study of ancient Roman construction and spent several years in Rome, where bridges with iron clamps securing the blocks of masonry (see Section 3.9) were still to be seen and in use. It is known that he built several models and he may have tested some to destruction; certainly he could have determined the critical angle of friction by experiment. Parsons (Ref. 7.15) believes that he obtained his design from observations of building failures, both ancient and modern.

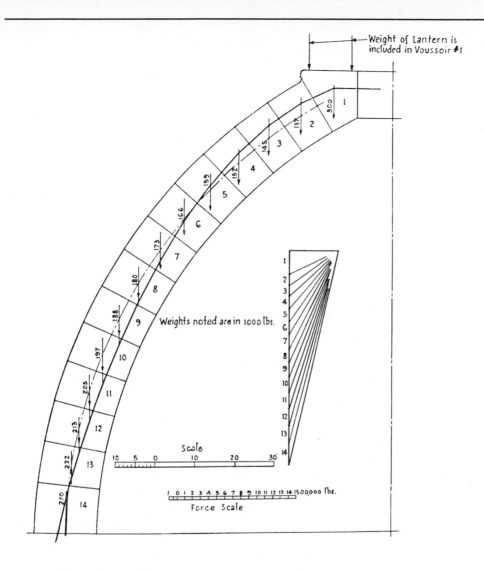

Weight of Lantern is included in Voussoir #1

Weights noted are in 1000 lbs.

Scale

Force Scale

7.3

Line of thrust in one of the main ribs supporting the dome and the lantern, as drawn by Parsons (Ref. 7.15, p. 597).

The dome was started in 1420 and finished in 1434. No masonry dome with a greater span has been built since that time. Table 7.1 is a list of the masonry domes built with a span of more than 30 m (100 ft) and with no or only a small amount of reinforcement.

The dome of S. Maria del Fiore has, like the Pantheon, never given subsequent cause for concern about its safety, and in this respect it differs from most Gothic structures and from S. Pietro in Rome.

Table 7.1

Name	City	Date Completed	Span m	Span ft
Pantheon	Rome	A.D. 123	43	143
S. Sophia	Istanbul	537	33	107
S. Maria del Fiore	Florence	1434	42	138
S. Pietro	Rome	1590	42	137
Gol Gomuz	Bijapur, India	1656	42	137
St. Paul's	London	1710	33	109
Mosta Church	Mosta, Malta	1840	38	124
S. Carlo	Milan	1847	32	105

Not everything Brunelleschi undertook was a practical success. In 1430 he was sent as military engineer to Lucca, taking Michelozzo, Donatello, and Ghiberti with him as his assistants. He decided to turn the river Serchio into the defensive trenches by means of a canal in order to cut off the city and force its capitulation. The diversion was a complete success, and Lucca was converted into an unapproachable fortress in the middle of a great lake. The Florentine forces withdrew without taking Lucca (Ref. 7.15, p. 600).

7.3   ST. PAUL'S CATHEDRAL

In chronological order, the next great dome of the Renaissance was S. Pietro's in Rome, but because it gave rise to the first major discussion of structural theory we shall consider it in Section 7.6.

The Renaissance reached England slowly during the reigns of Henry VIII and Elizabeth I. In 1612 Inigo Jones, who had the year before become Surveyor of the King's Works, paid his second visit to Italy. He was particularly impressed by the works of Andrea Palladio (see Section 7.10) and on his return introduced Palladian principles of design into English architecture.

Sir Christopher Wren, sixty years younger than Inigo Jones, was appointed Professor of Astronomy at Gresham College, London, in 1597 at the age of 25 and Savilian Professor of Astronomy at Oxford in 1661. Like many Renaissance scientists, he had an amateur interest in architecture and was the designer of the Sheldonian Theatre still used for ceremonial occasions by Oxford University. Following the Great Fire of London in 1666 (see Section 7.8), he was appointed Surveyor of Works (1669). Wren was unable to obtain authority for major replanning of the city and the pattern of streets, deriving largely from Roman times, was retained (see also Section 3.8). He did, however, rebuild the churches destroyed in the fire in the Renaissance style.

The old St. Paul's Cathedral had a Norman nave and an Early English (i.e., early Gothic) central tower. Inigo Jones had added a classical portico. The cathedral had given cause for concern for at least two centuries before the fire, for the nave was out of the perpendicular and the masonry of the central tower was badly cracked. Inigo Jones had started a program of restoration during the reign of Charles I but this had been interrupted by the Civil War. Wren had suggested a major restoration after his visit to Paris in 1665 (Ref. 7.3a, p. 43), and his proposal included the removal of the central tower and its replacement with a great dome.

The cathedral was not wholly destroyed by the fire and the original intention was to repair the nave, but during the restoration in 1668 a collapse occurred and it was then decided to build a completely new cathedral. The rebuilding of St. Paul's has been described in several books (e.g., Ref. 7.17) and is fully documented in the publications of the Wren Society (Ref. 7.3). These volumes include the correspondence between W. Sancroft, the Dean of St. Paul's, and Wren, following the collapse of 1668, various studies and models made by Wren for the design of the dome, detailed accounts of the cost, and the minute book of H. M. Commission for Rebuilding St. Paul's Cathedral. We have no details of Wren's structural design.

The solution adopted to support the unusually tall dome with its heavy lantern is most ingenious (Fig. 7.4). The dome seen from the inside carries merely its own weight. Above it a brick cone tied with iron chains at the base supports the 700-tonne (700-ton) masonry lantern, its own weight, and, in part, the timber trusses that carry the lead-covered outer timber dome. The outer dome gives the impression of a masonry structure, an illusion conveyed by the masonry lantern and the lead covering over the timber (Fig. 7.5); however, it would not be possible to build a masonry dome with this ratio of height to span without buttresses or a great deal of reinforcement.

As we now know, the structurally most advantageous shape for a concentrated load is a straight cone; for a weight distributed uniformly over the surface the most advantageous form is a catenary-shaped cone. Wren's cone is near-linear but with a slight swelling to allow for its own weight and the reactions of the timber trusses. A century later the correct shape was being determined by observing the curve adopted by an appropriately loaded cable hanging under its own weight (see Section 7.6). If Wren was already familiar with this simple experimental technique, he does not mention it.

The construction of the dome of St. Paul's was begun in 1697. Robert Hooke, who was closely associated with Wren as curator of the Royal Society and as one of the City Surveyors, is reported in the Minutes of the Royal Society for 1670 to have "brought in this problem in architecture—the basis of the pillars and the altitude of the arch being given, to find out the right figure for that arch." Asked for a demonstration, he said that he would show it to the President. The matter was raised again in 1671; Hooke replied that he had shown it to the President but the minutes give no details.

In 1675 he wrote a series of anagrams in his book on helioscopes "to fill a vacancy on the ensuing space." One of these was deciphered by Richard Waller in 1705 to mean, "As hangs the flexible line, so but inverted will stand the rigid arch" (Ref. 7.4, p. 4).

In 1697 David Gregory published in the *Philosophical Transactions of the Royal Society* a paper *On the properties of the Catenaria*, which stated that the theoretically correct line for an arch is an inverted catenary:

And when an arch of any other figure is supported, it is because in its thickness some catenaria can be included. Neither would it be sustained if it were very thin, and composed of slippery parts . . . .

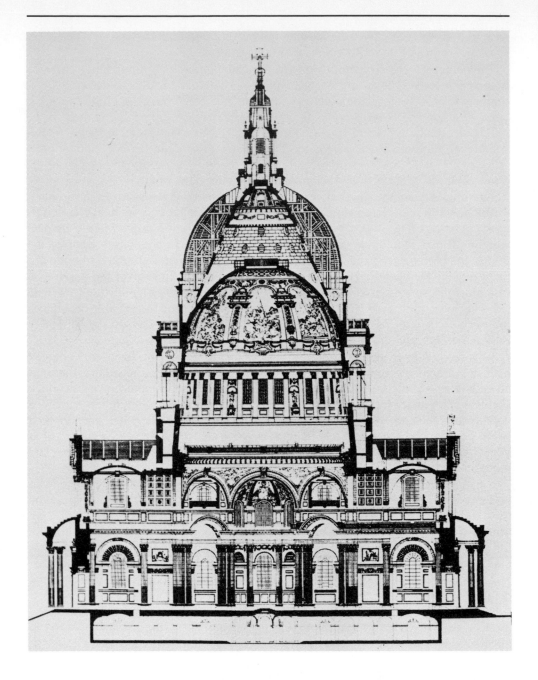

7.4

Section through St. Paul's Cathedral, taken from a drawing by A.F.E. Poley.

7.5

The dome of St. Paul's Cathedral, London, was modeled on the earlier dome of S. Pietro (Fig. 7.13); however, the outer dome was actually built as a lead-covered timber frame. The resemblance is purely visual.

For the force, which in the chain draws inwards, in an arch equal to the chain drives outwards. All other circumstances, concerning the strength of walls to which arches are applied, may be geometrically determined from this theory, which are the chief things in the construction of edifices (Ref. 7.16, p. 76).

The first sentence in this statement is given without proof or reason and was, in fact, not proved until 1776 by Coulomb (see Section 7.6). How Gregory obtained this important result we do not know; however, it is quite correct.

Gregory had in 1691 been appointed to the Savilian Chair of Astronomy at Oxford, occupied by Wren until 1673. In 1692 he was elected a Fellow of the Royal Society, of which Wren became President in 1681. New ideas do not appear suddenly, and Wren and Gregory (who was twenty-nine years younger) belonged to the same small scientific circle. Hooke was professionally close to Wren, although inclined to be secretive. It is therefore probable that Wren was familiar with the concept that the correct shape of an arch could be obtained from that adopted by a string carrying the same load. Of all the great architects of the Renaissance Wren had the best scientific training. Robert Hooke, who was a greater scientist, said of him:

I must affirm, that since the time of Archimedes there scarce ever has met in one man, in so great perfection, such a mechanical hand, and so philosophical a mind (Ref. 7.17, p. 134).

Another possible source of inspiration is the roof of the Baptistry in Pisa (Ref. 7.13, plates 118–123). Although Wren never visited Italy, he may have heard of the Baptistry from his friend John Evelyn who had studied in Padua. This dome, 39 m (127 ft) in diameter and pierced by a cone, was completed in 1278. The cone is a little steeper than Wren's, carries no lantern, and is actually part of the visible outer roof of the Baptistry. We know nothing of its design.

The use of a cone hidden between the inner and the outer domes has been attacked, both during the Neo-Gothic era and in the midtwentieth century, as structurally dishonest. This argument is reinforced by Wren's use of flying buttresses, hidden behind a screen wall, to convey the horizontal thrust of the vault over the nave to the ground through the outer walls. Like Brunelleschi, Wren was determined not to allow Gothic buttresses to be seen or to allow his interior space to be interrupted by exposed iron ties. He also wanted a dome with a greater ratio of height to span than either that of the Duomo of Florence or of S. Pietro in Rome. This may have been conditioned by the restricted site that limited the angles from which the dome could be seen. It is difficult to determine how Wren could have achieved these aims, given the technology of his time, without the cone. Certainly Wren's design was the most economical long-span structure before the nineteenth century, for the loadbearing brick cone is only 18 in. (450 mm) thick.

The quality of the workmanship, however, was less satisfactory. It is possible that Wren, having little practical experience of building, left too much to his master masons Marshall and Strong (Ref. 7.3c, p. ix), who had been trained in the ancient traditions of medieval masonry. Thus the great pillars supporting the dome were built in the Gothic manner with a core of rubble, apparently derived from the demolition of the old St. Paul's, and only the outside veneer was of Portland stone. The bursting pressure in this veneer became evident in 1695, and in the building accounts for December 1709 there appears an item:

Repairing flaws occasioned by the pressure, making good &c the 8 leggs of the Dome and inside E. W. N. & S. Walls of N. and S. Cross. Masons 420 days. Labourers 289 &c £1,283.8.11 (Ref. 7.3c, p. x).

*7.4   THEORETICAL MECHANICS IN THE RENAISSANCE

The effect of the Renaissance on practical construction was immediately evident. Brunelleschi's dome was one of the earliest achievements of the Renaissance and is still considered one of its greatest architectural monuments. The foundations of structural mechanics belong to the same period, but it took another three centuries before they began to exert a visible influence on the design of buildings.

Science advanced slowly during the fifteenth and sixteenth centuries, but by the seventeenth it was far ahead of Hellenistic science. One important reason was the greater willingness of Renaissance scientists to experiment.

Virtually every aspect of structural mechanics received some attention from Leonardo da Vinci, the most versatile man of the Renaissance, who was born near Florence in 1452 and died in France in 1519. Leonardo published nothing during his lifetime, but he left his extensive notes to his friend and pupil Francesco Melzi, after whose death the notebooks were dispersed; some have been lost. Leonardo, who was left-handed, wrote in a hand that ordinary persons can read only in a mirror, and his abbreviations were many. The illustrations have been known for a long time, but most of the writing was not transcribed until the twentieth century. The effect of Leonardo's work on the development of mechanics was therefore barely felt; however, his ideas are of interest, for they show how these problems were considered by the only scientist of the fifteenth century whose detailed notes survive.

Leonardo has been described as a great scientific genius by many commentators. There are others (e.g., Truesdell, Ref. 7.18) who dismiss his contribution as negligible. Leonardo's critics are correct in claiming that his work was unsystematic, that he recorded observations and *suggested* experiments but did not describe those that he actually carried out, and that some of his definitions, although important-sounding were meaningless; to illustrate, among his many similar formulations of force (*forza*) the following may serve as an example:

Force I define as an incorporeal agency, an invisible power, which by means of unforeseen external pressure is caused by the movement stored up and diffused within bodies which are withheld and turned aside from their natural uses; imparting to these an active life of marvellous power it constrains all created things to change of form and position, and hastens furiously to its desired death, changing as it goes according to circumstances. When it is slow its strength is increased, and speed enfeebles it. It is born in violence and dies in liberty; and the greater it is, the more quickly is it consumed. It drives away in fury whatever opposes its destruction. It desires to conquer and slay the cause of opposition, and in conquering it destroys itself." (From Manuscript A in the Library of the Institut de France, folio 53, verso, Ref. 7.5, Vol. I, p. 493).

* This section assumes some knowledge of mechanics. It can be omitted without loss of continuity of the story.

This may be good poetry, but as a scientific definition it is no more useful than the notion common in the early Middle Ages that a stone falls to the ground because it is of the earth and therefore returns to it when given the opportunity.

The definition of force as a quantity with both magnitude and direction is fundamental to any theory of structural mechanics. Leonardo never put this into words, but it was implicit in many of his diagrams and comments, and this was the first time that the composition of forces was usefully described.

Medieval scholars got no further than a solution of the law of the inclined plane, credited to Jordanus Nemorarius, a thirteenth-century monk about whom little is known except the manuscript he left. The inclined plane was considered in terms of diluted gravity (*gravitus secundum situm*)—the flatter the plane the smaller the *gravitas secundum situm*.

Leonardo dealt with the condition of equilibrium of two inclined strings carrying a weight (Fig. 7.6). He noted elsewhere (Ms E in the Institut de France, f. 60 v) that the

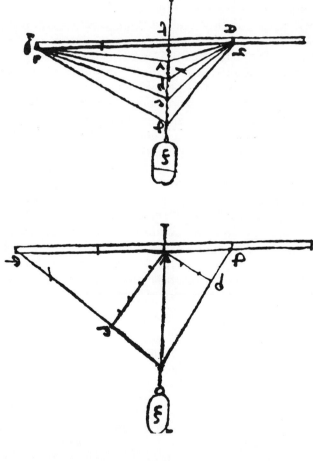

7.6

Composition of forces, from the Arundel Manuscript, No. 263, in the British Museum, folio 1, verso.

The Renaissance

pull in a string could be ascertained by hanging it over a pulley and balancing it with a weight. If one then hung a weight from two inclined strings and set out the pull in the strings and the weight carried by them to scale, a parallelogram would be obtained. This is the *parallelogram of forces* which, together with Archimedes' lever principle (see Section 2.3), forms the basis of statics. It was rediscovered by Simon Stevin of Bruges (Belgium) in 1586. There is no reason to believe that Stevin (or even Galileo) knew what was in Leonardo's notebooks.

Even more prescient was Leonardo's solution of the breaking strength of an arch (Fig. 7.7):

The arch will not break if the chord of the outer arch does not touch the inner arch. This is manifest by experience, because whenever the chord *aon* of the outer arch *nra* approaches the inner arch *xby,* the arch will be weak, and it will be weak in proportion as the inner arch passes beyond the chord (Ms A in the Institut de France, f. 51 recto).

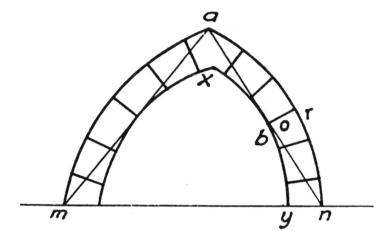

7.7

Leonardo's diagram for the breaking strength of an arch loaded at the crown, redrawn by Parsons (Ref. 7.15, p. 71).

This conclusion was based on the application of a single concentrated force at the crown of the arch, a simpler problem than that posed in Gothic architecture in which the load was distributed. The masonry arch is stable as long as the line of thrust falls wholly within the cross section. This result was rediscovered by P. Couplet in a paper published in Paris in 1730 (Ref. 7.16, Fig. 6.6, p. 171). These two results would have been sufficient for the design of a large part of Gothic and Renaissance architecture.

Leonardo also anticipated the testing of materials, which is a basic element of modern engineered construction (Fig. 7.8):

The object of this test is to find the load an iron wire can carry. Attach an iron wire, 2 braccia (1.2 m or 4 ft) long to something that will firmly support it, then attach a basket or any similar container to the wire and feed into the basket some fine sand through a small hole placed at the end of a hopper. A spring is fixed so that it will close the hole as soon as the wire breaks.

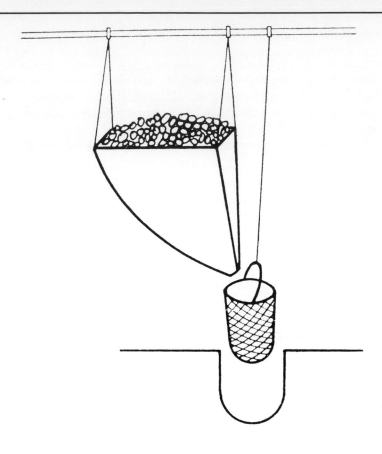

7.8

Leonardo's diagram for the tensile testing of wires, redrawn by Parsons (Ref. 7.15, p. 72).

The basket is not upset while falling, since it falls through a very short distance. The weight of the sand and the location of the fracture are to be recorded. The test is repeated several times to check the result (Arundel Manuscript, folio 82, verso).

The machine suggested is similar to that designed by Michaelis for testing cement briquettes in tension, used extensively in Germany and England and still found in some laboratories (Ref. 7.6, Fig. 239, p. 535). The suggestion that the test should be repeated also accords with the modern observation that the strength of a material varies slightly from one piece to another.

Leonardo failed to obtain a solution for the all-important problem of the beam. He came back to it many times (Ref. 7.18, pp. 13–20) and established that the load-carrying capacity was directly proportional to the width and inversely proportional to the span. He did not mention its relation to the depth of the beam, possibly because he was unable to express a quadratic relationship. The mathematics in Leonardo's notebooks is, by the standards of his time, primitive and all relationships quoted by him are linear.

It is puzzling that the eminently useful results obtained by Leonardo had no effect on theoretical mechanics. He does not appear to have been unduly secretive and he was certainly not modest. In a letter to Ludovico Sforza, Duke of Milan, he offered to make "extremely light and strong bridges"; others "secure and indestructible by fire and battle"; and "methods of destroying those of the enemy." He also offered to design "mortars, most convenient and easy to carry" and "big guns." Leonardo's mechanics were unsystematic and without proper proof or experimental verification. They were also perhaps ahead of their time. The fifteenth and sixteenth centuries were notable for their contribution to art and architecture but not to science and mathematics; the scientific Renaissance began only at the end of the sixteenth century. The notebooks are interesting, however, for they showed what could be done by one man in the later fifteenth century, although there may have been others whose work is lost.

As noted earlier in this section, Stevin was the first to prove formally the theorem of the composition of forces, published in a Flemish-language book *De Beghinselen der Weeghconst* (elements of statics) in 1586. We also noted that Leonardo used weighted strings to draw the parallelogram, a method still used in student experiments today. Stevin's derivation was based on medieval orthodoxy. The problem of the inclined plane had been solved by Jordanus Nemorarius. Stevin thus considered a *clootcrans* (literally a wreath of spheres, a continuous chain of spheres) moving over a triangle with two inclined planes (Ref. 2.6, Vol. I, p. 167). These must be in equilibrium or perpetual motion would result. From this he deduced the equilibrium of *two* spheres joined by a string on two inclined planes and obtained the triangle of forces (which is half the parallelogram of forces and sufficient for the solution).

Architectural historians generally terminate the Italian Renaissance in the midsixteenth century; they call the work of the late sixteenth Mannerism and that of the seventeenth Baroque. In the history of science these words have no meaning, and if the term Renaissance is used at all it is meant to extend to the Age of Reason. Thus Galileo is usually considered a scientist of the Renaissance.

Galileo Galilei is best known for his work on celestial mechanics. His defense of Copernicus' theory (that the earth moves round the sun and not the other way round) brought him into conflict with the Inquisition. In spite of his friendship with Pope Urban VIII (formerly Cardinal Maffeo Barberini), to whom he had dedicated one of his books, he was forced to confine himself to terrestrial studies. This may have been fortunate for structural mechanics, for while under house arrest Galileo produced the first book that dealt with its theory, *Due Nuove Scienze* (Ref. 7.19). This work was written in Italian, but published in Leiden in 1638, because at that time the Inquisition refused permission for the publication of any book by Galileo, and Holland was outside its control. It was reprinted in Bologna in 1655.

In this book Galileo produced the first usable formula for the strength of a beam. Taking a cantilever (Fig. 7.9), he assumed that the beam would separate from the support AB, for he had probably observed many flexural tension failures in masonry cantilevers. The moment of the tensile force in the beam therefore had to balance the bending moment about the point of rotation B. There was no question of locating the

7.9

Illustration of the bending test (from Galileo Galilei, *Due Nove Scienze,* Elzevir, Leiden, 1638).

neutral axis or determining the stress distribution; these are matters that were not thought of until the eighteenth century (see Section 8.3). Galileo described his theorem:

It is clear that, if the cylinder breaks, fracture will occur at the point B where the edge of the mortice act as a fulcrum for the lever BC, to which the force is applied; the thickness of the solid, BA, is the other arm of the lever along which is located the resistance. The resistance opposed the separation of the part BD lying outside the wall, from the portion lying inside. From the preceding it follows that the magnitude of the force applied at C bears to the magnitude of the resistance the same ratio, which half the length BA bears to the length BC.

Thus Galileo obtained quite correctly that the load-carrying capacity is proportional to the width, to the square of the depth, and inversely proportional to the span. Using modern terminology (see Section 8.3)

$$M = \tfrac{1}{2}fbd^2 \qquad (7.1)$$

whereas the correct elastic answer is $M = \tfrac{1}{6}fbd^2$. If the formula is also used to *determine* the comparative strength of the material f, the numerical error disappears,

The Renaissance

but we have no evidence that this formula was ever used in the design of an architectural structure. The correct solution was published by Navier in 1826 (see Section 8.3).

Using his theory of bending, Galileo then established the rules of physical similitude and thus laid the theoretical foundations for model analysis (Ref. 7.7). He noted that the load that can be carried by a beam of square cross section $d \times d$ and length L is proportional to d^3/L, from which he derived his law of the "weakness of giants":

If one wished to keep, in an immense giant, the proportions of the members of a normal man, it would be necessary to find a much harder and stronger material.

If the weight of the giant structure is W_2 and the weight of the normal structure is W_1, their respective sizes are L_2 and L_1, and the strength of the materials is, respectively, f_2 and f_1, then

$$\frac{W_2}{L_2{}^2 f_2} = \frac{W_1}{L_1{}^2 f_1} \tag{7.2}$$

This relationship can be used to determine how soon a structure will collapse under its own weight or how big a man can be without breaking his bones under his own weight (although the latter problem is rather more complicated than Galileo supposed). Another application to this law of similitude was not mentioned by Galileo and apparently was overlooked until the nineteenth century (see Ref. 1.3, Section 5.2). If a model of a complex structure is made, say, to a scale of 1:6 or 1:10, then the test load will indicate the load-carrying capacity of the full-size structure in accordance with (7.2). This could have been used with great advantage for architectural structures during the seventeenth and eighteenth centuries.

7.5 THE HANDLING OF GREAT WEIGHTS

The Renaissance had neither the manpower nor the skill of ancient Rome for handling materials, but it improved on medieval procedures (see Sections 2.4 and 6.4). Sophisticated compound tackles of the type described by Hero and known to have been used in Roman times (Figs. 2.4 and 2.5) came back into use, and hoisting gear capable of moving material both horizontally and vertically was invented (Fig. 7.10). The restriction on horizontal movement had been one of the limitations suffered by the medieval builder.

The difficulties experienced by Renaissance engineers are best illustrated by the erection of the obelisk in the Piazza S. Pietro in 1586. Camillo Agrippa, who made one of the proposals for moving the obelisk, stated that the project was already under consideration when he arrived in Rome in 1535. It may well have been the biggest material-handling operation in a thousand years and thus caused a long public debate.

7.10

Revolving crane, from *Diverse et Artificiose Machine* by Agostino Ramalli, published in 1588.

7.11

Erection of the Vatican Obelisk in the Piazza S. Pietro, from Fontana's
Della trasportatione dell'obelisco Vaticano.

We are particularly well informed about this matter because Domenico Fontana, who planned and supervised the move, was the author of *Della trasportatione dell'obelisco Vaticano et delle fabriche di nostro Signore Papa Sisto V, fatte del Cavallier Domenico Fontana, architetto di Sua Santità,* a lavishly produced and heavily illustrated folio of 108 pages, published in Rome in 1590. It was probably intended to be presented to potential noble clients.

When medieval S. Pietro was replaced by the present much larger cathedral, the obelisk stood in the way of the development of the site. The decision was then made to move it from the back to the front of the church. The obelisk had been cut from hard rock in Upper Egypt, in the tenth century B.C. and transported a great distance before being erected. The Romans had removed it to Rome in A.D. 41, again transporting it a long distance and raising it in an area subsequently called the Circus of Nero. On this latest occasion it was necessary to move the obelisk a distance of only 115 canne (257 m or 843 ft).

Fontana measured the obelisk, obtained a sample of a *similar* stone, and calculated its weight as 963,537 35/48 libbre (328 tonnes or 361 U.S. tons). This meaningless precision was not uncommon at that time. He then estimated that a windlass worked by horses could lift 20,000 libbre and thus obtained the number of tackles required. These, however, were the only calculations. The triangle of forces had not yet been published, and another century passed before it was used to solve architectural problems.

In order to convince the committee of the soundness of his scheme, Fontana made a scale model of timber with a lead obelisk and demonstrated the technique. He was then given sweeping powers to requisition timber, rope, victuals, horses, and workmen, with the proviso that due payment would be made. The actual operation was undertaken with 40 windlasses, 140 horses, and 907 men. On the preceding day the workmen were confessed and given the sacrament, and two masses were read before dawn on the day the obelisk was to be lowered. Troops were posted to restrain the crowds, and orders were issued that anybody who spoke or made a loud noise would be subject to the death penalty; a public executioner is reported to have been present on the site. The various operations were carefully prearranged and coordinated by trumpet signals, and the obelisk was lowered in one day, but the move of 250 m (800 ft) along a special causeway and preparation for the erection took six months. Hoisting into a vertical position was then accomplished in another day between dawn and dusk (Fig. 7.11).

7.6 THE DOME OF S. PIETRO IN ROME AND THE THEORY OF ARCHES AND DOMES

The old Basilica of S. Pietro was built in A.D. 330 in the Circus of Nero, the site of the martyrdom of St. Peter. The original design of the new basilica was the work of Donato Bramante and the foundation stone was laid in 1506. The direction of the construction then passed successively to Raphael, Baldassare Peruzzi, Antonio da Sangallo the Younger, and Michelangelo Buonarotti. Michelangelo planned the dome and completed the drum. On his death in 1564 he was succeeded by Vignola, but he left drawings and models for its completion up to the lantern. In 1585 Giacomo della Porta and Domenico Fontana (who later moved the Vatican Obelisk; see Section 7.5) began the construction of the dome. The building was completed by Carlo Maderna in 1612.

The dome is a little smaller in diameter than that of the Duomo in Florence (41.9 m or 137.5 ft), but the overall height of the basilica is greater. The top of the cross is 138 m (452 ft) above the ground, which is unsurpassed by any Renaissance church and among Gothic cathedrals only by Strasbourg; the spire of Ulm Cathedral (161 m or 529 ft) dates from the nineteenth century.

Bramante's design (Ref. 7.15, Fig. 210, p. 608) was conservative. He proposed a solid hemispherical dome, thickened in its lower portion, with steps showing outside as in the Pantheon (see Section 3.9), clearly its inspiration. The drum of the dome was to be supported on four semicircular arches which in turn were to rest on four great pillars. The design was less efficient than Brunelleschi's, which had a double masonry shell and pointed arches. The use of a circular instead of an octagonal plan was a structural advantage, although probably neither of them knew it.

Bramante built the four piers after having experimented with Roman methods of concreting. The piers with their concrete core are now inside the enlarged piers built subsequently. On these piers he raised the four great arches, 46 m (150 ft) high and spanning 26 m (84 ft), the biggest since antiquity.

Raphael, who was Bramante's nephew, succeeded him, but he had no structural experience and added nothing to the structure. Sangallo the Younger strengthened the pillars left by Bramante and built the vault over the nave, using scaffolding similar to that employed by Brunelleschi (Fig. 7.1); views of the vault with the scaffolding in place are shown in a number of contemporary prints and in a painting by Giorgio Vasari. Sangallo also built the pendentives that define the drum of the dome. He redesigned the dome (Ref. 7.15, p. 609) by changing the hemispherical form to a segmental arc of revolution and increasing the height by 9 m (30 ft), but he omitted the stepped-up rings that Bramante had copied from the Pantheon and substituted longitudinal ribs.

The design of the dome as actually built is largely Michelangelo's, who is best known as a great sculptor but had in middle age distinguished himself as a military engineer when in 1529 Florence was under attack. At the time of his appointment as the architect of S. Pietro he was 72 years old.

Michelangelo wrote to Florence and asked for the dimensions of the dome. He changed the design from a solid to the more efficient double-shelled structure devised by Brunelleschi, returned the shape to Bramante's hemisphere, and increased the number of longitudinal ribs to sixteen. He made a number of models of his design, of which a large wooden model survives, as have a number of his drawings, listed by Ackerman (Ref. 7.20, p. 333). Having decided on the shape of the dome, Michelangelo concluded that Sangallo's pillars were inadequate and he strengthened them further.

Twenty-five years after Michelangelo's death Giacomo della Porta elevated the height of the dome and abandoned the hemispherical shape in favor of that used by Brunelleschi and proposed by Sangallo, although some critics (Ref. 7.20) attribute this change to Michelangelo. He reduced the thickness of the external ribs and lowered the lantern. The surviving wooden model is actually Michelangelo's modified by della Porta. The construction of the dome was completed during the next five years (1585–1590), with Giacomo della Porta and Domenico Fontana as joint architects (Fig. 7.12). Contemporary views of the scaffolding show that it was similar to Brunelleschi's (Fig. 7.1).

7.12

The dome of S. Pietro, Rome, a brick dome covered with lead.

It is not known how much tension reinforcement was used in the dome and who designed it, but there are three iron chains around it. It has been suggested from time to time that iron clamps join the blocks of masonry (e.g., Ref. 7.14, p. 58). The predominant material of the dome is brick.

The tension reinforcement was insufficient because the dome subsequently showed alarming cracks which led to a number of investigations of its structural behavior. These are the first examples of the application of structural mechanics to an architectural problem and we consider one of them in some detail.

We have already noted (Section 7.3) Gregory's paper *On the properties of the Catenaria*, published in 1697. Two years before Philippe de La Hire, Professor of Mathematics at the *Académie Roïale des Sciences* in Paris, published *Traité de Méchanique* (A Treatise on Mechanics) in which he argued that the shape of the arch must be such that for each block the resultant of its own weight and of the pressure of the preceding block is perpendicular to the face of the next block (Fig. 7.13). The arch is then stable even if no friction occurs between the joints. The significance of this assumption becomes clear if the wedge-shaped blocks are replaced with circular spheres which can roll over one another (Fig. 7.14a). In the absence of friction the spheres behave exactly the same as the wedge-shaped blocks, provided they have the same weight. De La Hire's theory is therefore identical to Gregory's and was one of those used for checking the safety of the dome of S. Pietro. By 1740, that is, 150 years after the dome was completed, some of the cracks, which are always present in masonry, had widened so much that concern was felt for its safety. A number of experts were consulted in 1742 and 1743, one of whom was Giovanni Poleni, Professor of Experimental Philosophy (which was roughly equivalent to physics) at the University of Padua. He submitted his report in 1743 and published it in book form in 1748. His theory was based on *Lineae Tortii Ordinis Neutoneanae*, a book

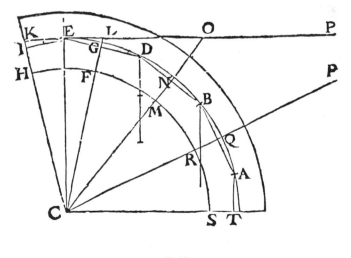

7.13

Phillippe de La Hire's solution to the masonry arch, from *Traité de Méchanique*, Paris 1729 (a later edition).

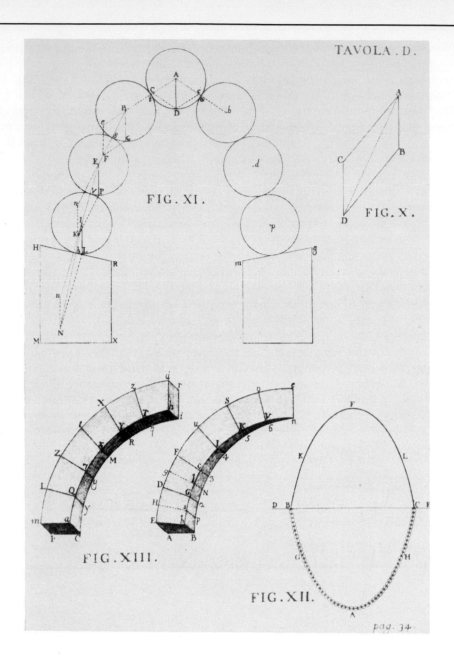

7.14a

The parallelogram of forces (X): the "smooth-sphere" analogy for the masonry arch (XI); the use of the catenary for the solution of the masonry arch (XII and XIII) from *Memorie istoriche della Gran Cupola del Tempio Vaticano* by Giovanni Poleni, Padua, 1748.

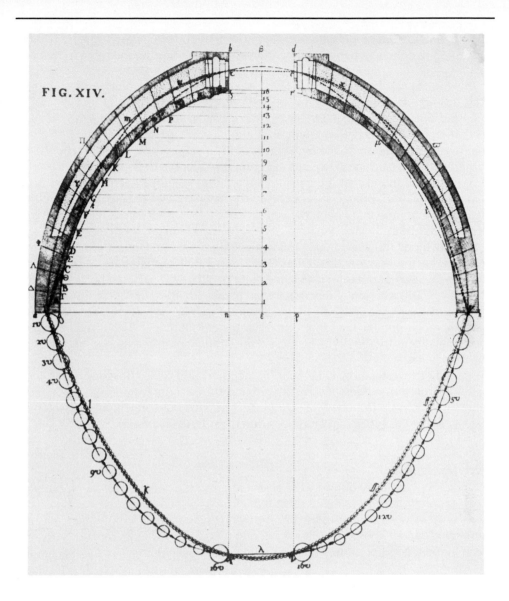

7.14b

The catenary chain for a dome of uniform weight and for the dome of S. Pietro; the line of thrust represented by the catenary superimposed on the cross section of the dome (from *Memorie istoriche della Gran Cupola del Tempio Vaticano* by Giovanni Poleni, Padua, 1748).

published in Oxford in 1717 by J. Stirling, which included Gregory's theory with the suggestion that the catenary might be formed by balancing smooth (i.e., frictionless) spheres. This suggestion is reproduced by Poleni (Fig. 7.14*b*).

The vertical cracks in the dome divide it into a series of "orange-slice" segments, each forming an arch. Provided every one of these arches is safe, the cracks are harmless. Poleni therefore made up a string of beads, the weight of each bead being proportional to the weight of a unit portion of a standard orange-segment. He then determined its catenary shape by hanging the string from two supports (Fig. 7.14*b*). Turning this curve upside down and superimposing it on a cross section of the dome of S. Pietro, he found that the catenary lay entirely within the cross section of the dome and that the "orange-slice arches" were therefore quite safe. It is interesting, incidentally, to note that Poleni derived his parallelogram of forces (Fig. 7.14*a*) from Newton, not from Stevin.

He concluded, however, again following Gregory, that adequate provision had to be made for the horizontal reaction. The inward pull of the string of beads could easily be measured and from it an assessment could be made of the horizontal thrust of the arch. This horizontal reaction had to be absorbed by the hoop stresses within the masonry (see Fig. 3.30) or by ties around the masonry dome. The dome of S. Pietro is much thinner than that of the Duomo in Florence. Poleni recommended additional chains for the dome and tested the strength of iron as part of his investigation to determine how much was needed.

In another investigation T. Le Seur, F. Jacquier, and R. G. Boscovich, the last a well-known Jesuit mathematician, arrived at a similar conclusion but with a higher value for deficiency in horizontal resistance. The method is ingenious but not in the mainstream of this history and has been described in detail elsewhere (Ref. 3.36, pp. 111–117).

In 1743 and 1744 five additional tie rings were added to the dome of S. Pietro by Luigo Vanvitelli.

Many more masonry domes were designed up to the twentieth century, but none raised problems as interesting as the three we have discussed and all were designed more or less by empirical rules, for the theory was perfected only after masonry domes had become obsolete. Masonry domes therefore now fade from our story, and this may be a good point to summarize the further development of the mechanical theory.

A. A. H. Danyzy tested small plaster arches at the Academy of Montpellier, in southern France, and observed that failure commonly occurred by rotation of the voussoirs (Fig. 7.15). He presented these results at a meeting of the *Société Royale* of Montpellier in 1732 and eventually published them in 1778 (Ref. 7.16, p. 201). A masonry arch is a rigid frame and consequently has three redundancies (see Ref. 1.3, Section 3.3). Three "hinges" are needed to turn the statically indeterminate arch into a statically determinate structure, a condition that is critical for its safety; a fourth hinge, shown in Fig. 7.15, turns the arch into a mechanism and causes collapse.

Charles Augustin de Coulomb solved this problem in a paper written while he was officer in charge of public works on the French West Indian island of Martinique. Quite apart from the language problem, this is difficult for the modern reader to understand, and it has been translated and published by Heyman with an admirable commentary (Ref. 7.16).

7.15

A model of lead or other heavy voussoirs (segmental blocks) shows the failure of the masonry arch by the formation of four hinges. Three hinges make the rigid arch statically determinate; four turn it into a mechanism and cause collapse (*Architectural Science Laboratory, University of Sydney*).

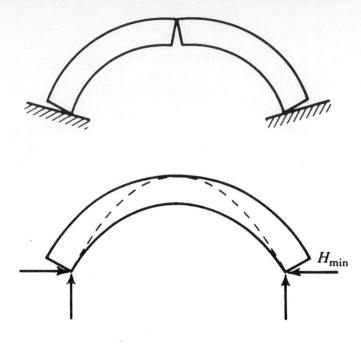

7.16a

Coulomb's theory illustrated by Heyman. The hinges occur where the line of thrust touches the intrados or the extrados of the arch. The arch is just stable with three hinges. (i) Line of thrust for arch carrying a vertical load (including its weight).

Coulomb distinguished between the line of thrust produced by the weight of the arch and a vertical load carried by the arch and the line of thrust due to movement of the abutments (Fig. 7.16). "Hinges" from where the line touches the intrados or extrados of the arch, as observed by Danyzy. A further hinge causes collapse. The line of thrust must therefore lie wholly within the cross section of the arch, as Gregory surmised.

Because the shape of the catenary depends on the load, we can for any given shape and thickness of arch determine the ratio of thickness to span (t/L) needed to ensure stability of a masonry arch. Such simple proportional rules developed during the nineteenth century and were discussed in a number of architectural articles and textbooks (see also Section 7.11).

As demonstrated by Poleni, the vertical (meridional) forces can be determined by the theory of the masonry arch, and, as we noted in Fig. 3.30, the horizontal (hoop) forces change from tension to compression at an angle of 52° 24′ with the crown.

Thomas Young, in his lectures at the Royal Institution (Sections 8.3 and 8.6), stated that tension in a dome would develop only if the span exceeded 11/14 of the diameter of the sphere from which the dome was cut; this is equal to an angle of 51° 46′. He also made an interesting observation on the stability of the dome:

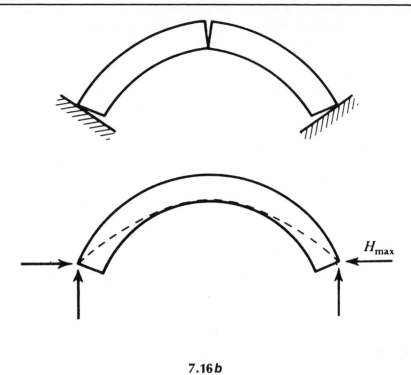

7.16 b

(ii) Line of thrust due to movement of the abutments.

The construction of the dome is less difficult than that of an arch since the tendency of each arch to fall is counteracted not only by the pressure of the parts above and below but also by the resistance of those which are situated on each side. A dome may therefore be erected without any temporary support like the centre which is required for the construction of an arch, and it may be left open at the summit without standing in need of a keystone, since the pressure of the lower parts is sufficiently resisted by the collateral parts of the same horizontal tier (*A Course of Lectures on Natural Philosophy and the Mechanical Arts,* 1807).

A rigorous proof, based on elementary statics, was given by Dr. Edmund Beckett Denison (who became Sir Edmund Beckett and later Lord Grimthorpe) in *On the Mathematical Theory of Domes,* which appeared in the *Memoirs of the Royal Institute of British Architects,* February 1871, pp. 81–115; Denison stated that it was the first time the theory had been given. (I am unable to say whether the claim is correct.) Empirically, the result had been known for some time. The limiting ratios for t/L given by Denison (now Beckett) in the article on *Domes* in the ninth edition of the *Encyclopaedia Britannica* (Ref. 4.6, Vol. VII, pp. 347–348) are similar to those calculated recently by Heyman (Ref. 5.16, Figs. 10 and 16, pp. 233 and 236). The ratio t/L is lower for a dome than for an arch because of the more favorable distribution of its load; Heyman considered that a ratio $t/L = 0.05$ is sufficient for a full hemispherical dome, that is, a 43-m (140-ft) diameter hemispherical dome requires a thickness of 2.1 m (7 ft), including the space between the two shells of a double-shelled dome.

As Denison pointed out (Ref. 4.6), the hoop forces can be resisted only by a great thickness of material, as in the Pantheon and the Indian Gol Gomuz, or by buttresses or metal ties. Denison therefore correctly predicted the demise of masonry domes in favor of those built with iron; he could not foresee the versatility of the recently invented reinforced concrete.

Although Denison gave formulas for calculating the hoop forces, a more satisfactory method is based on the membrane theory (see Ref. 1.3, Section 6.3).

The entire theory of masonry domes was recently revised by Heyman (Ref. 5.16) in accordance with limit design (see Ref. 1.3, Section 4.10).

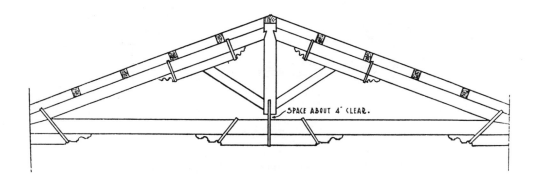

SPACE ABOUT 4" CLEAR.

7.17

Roof truss for the Uffizi Gallery in Florence, spanning 20 m (66 ft), designed by Vasari in the mid-16th century. The timber rafters are each in one piece, but there is a splice in the bottom chord at mid-span. There is only a nominal connection between the bottom chord and the king post (Ref. 7.15, p. 486).

7.7 TIMBER STRUCTURES

We noted in Section 5.6 that no authenticated medieval timber structures contain diagonal members. This lack of appreciation of the function of diagonals in trusses (see Ref. 1.3, Section 2.1) continued into the Renaissance. When bracing members were used, they were not always correct. Parsons (Ref. 7.15, p. 486) described roof trusses designed in the midsixteenth century by Giorgio Vasari for the Uffizi Gallery (Fig. 7.17). The diagonal members were not connected to bottom ties and were clearly conceived as stiffening for rafters, not as part of a truss.

Roman rules of truss design were revived by Andrea Palladio who, in his *Four Books,* gave this explanation:

Wooden bridges ought to be made in a such a manner, that they may be very strong, and tied together by large strong timbers, that there may not be any danger of their breaking, either thro' the great multitude of people, and of animals, or by the weight of the carriages and of the artillery that shall pass over them (Ref. 3.26, p. 63).

He then described a bridge of his own design over the Cimone River, between Trento and Bassano (Fig. 7.18):

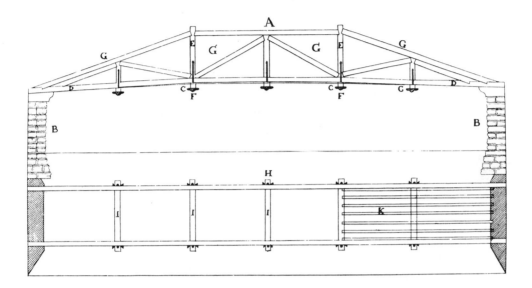

7.18

Palladio's bridge over the Cimone River, spanning about 30 m (100 ft). The annotation in the *Four Books* (Ref. 3.26, p. 65) is as follows: "A, The flank of the Bridge. B, The pilasters that are on the banks. C, The heads of the beams that form the breadth. D, The beams that form the sides. E, The colonelli. F, The heads of the cramps, with the iron bolts. G, Are the arms, which bearing contrary to each other, support the whole work. H, Is the plan of the bridge. I, Are the beams that form the breadth, and advance beyond the sides, near the holes made for the cramps. K, Are small beams that form the bed for the bridge."

The Cimone is a river, which falling from the mountains that divide Italy and Germany, runs into the Brenta, a little above Bassano. And because it is very rapid, and that by it the mountaineers send great quantities of timber down, a resolution was taken to make a bridge there, without fixing any posts in the water, as the beams that were fixed there were shaken and carried away by the violence of the current, and by the shock of stones and trees that by it are continuously carried down: therefore Count Giacomo Angaro, who owns the bridge, was under the necessity of renewing it every year.

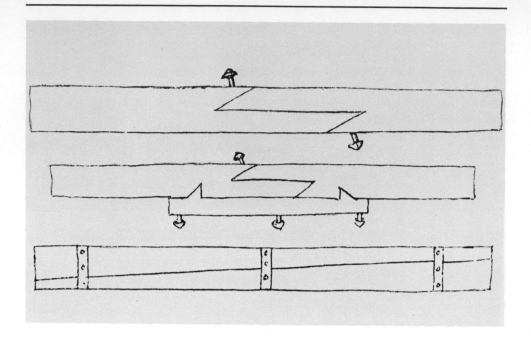

7.19

Joints in timber illustrated in a fourteenth-century manuscript by Taccola (Ref. 7.28, folio 46, verso).

The invention of this bridge is, in my opinion, very worthy of attention, as it may serve upon all occasions, in which the said difficulties shall occur; and because that bridges thus made, are strong, beautiful, and commodious: strong, because all their parts mutually support each other; beautiful, because the texture of the timbers is very agreeable, and commodious, being even in the same line with the remaining part of the street. The river where this bridge was ordered, is one hundred foot wide (Ref. 3.26, p. 65).

Although Palladio frequently invoked classical precedent, he did not do so in this case, and it may have been his own invention; however, the framing is similar to Apollodorus' bridge over the Danube (see Section 3.4) and two other bridges of his invention with crossed diagonals resemble it even more. It is noteworthy that Palladio described the diagonals G in Fig. 7.17 as "supporting the whole work." Timber structures with diagonal members became common after the sixteenth century, but Stevin's triangle of forces was not used for the design of trusses until 1847 (see Ref. 1.3, Section 2.1).

Timber joints capable of taking tension certainly existed in the fifteenth century (Fig. 7.19); it is possible that they are older in origin.

The Renaissance

7.8 BUILDING MATERIALS

In spite of the interest in Roman construction methods, the Renaissance did not revive Roman concrete, the most widely used material on Imperial Rome, except for some specialized applications (see Section 7.6). This may have been due in part to shortage of timber (for formwork), which was needed for shipbuilding and domestic construction, and was the principal source of domestic and industrial fuel. Clearing for agricultural land had already reduced the forests in medieval times. The developing metallurgical industries (iron, copper, and lead) and glass, salt, and gunpowder manufacture were also prime consumers of timber.

The lack of timber is emphasized by lengthy discussions during the Italian Renaissance of building vaults and domes without *armadura,* full timber formwork of the type used for modern vault and dome construction and for Roman concrete. The reason was not merely to save money but to conserve timber. One suggestion made in all seriousness for the construction of the cathedral in Florence (see Section 7.2) was to form the dome with earth mixed with coins; the poorer citizens could then be expected to remove the earth without payment in order to recover the coins (Ref. 7.15, p. 588).

The shortage of timber may also have been a reason for the increased demand for brick, which during medieval times had been employed mainly in those regions that lacked a supply of stone (Section 5.7). Loadbearing brick soon took the place of timber in domestic construction. The manufacture of bricks improved in the fifteenth century, and in 1625 English bricks were standardized by royal proclamation to 9 × 4 3/8 × 2 1/4 in. (229 × 111 × 57 mm).

The growth of the cities and the ever-present danger of fire may have been other reasons for the restrictions placed on the use of timber. In 1667, after the Great Fire of 1666, the British Parliament passed the London Building Act which prohibited timber-framed and half-timbered houses altogether and specified minimum thicknesses for external walls ranging from 22½ to 9 in. (572 to 229 mm).

Although most authenticated surviving medieval buildings are either churches or houses of the nobility, many domestic buildings in western Europe date from the fifteenth, sixteenth, and seventeenth centuries. In England, where the geological formations are sometimes no more than 40 km (25 miles) wide, a characteristic regional flavor was produced because building stone was generally obtained locally. In London, which is founded on clay, it was necessary to import stone; many of the prestige buildings (e.g., St. Paul's Cathedral) were built of a fine-grained, easily worked limestone quarried on Portland Isle, a peninsula in Dorsetshire, from which it was transported by water. In the nineteenth century Portland stone gave its name to the most common type of cement (see Section 8.7).

Wren is reported to have left his Portland stone to weather for three years to test its durability, a practice similar to that described by Vitruvius (see Section 4.1). It is quite likely that the practice of weathering stone and the regional character of masonry buildings go back to the Middle Ages, but data before the fifteenth century is lacking.

Glass was an expensive material in medieval times and had many flaws that limited its transparency. Clear glass had been made during the Roman Empire, but the technique was lost until its rediscovery in Venice in the fifteenth century. Venetian glass blowers brought it to England in the sixteenth.

The crown glass process, although a Syrian invention perfected in Venice, was especially developed in Renaissance England for use in windows. Crown glass was made by spinning (Fig. 7.20), which limited its size to about 54 in. (1.37 m) in diameter. Because it did not come into contact with any other material during manufacture it was free from defects which cast glass was not until the development of the Chance process (see Section 5.8). Crown glass dominated English window glazing until the nineteenth century.

The first record of crown glass windows in England is dated 1685, when Inigo Jones's Banqueting House in Whitehall was stripped of its original windows and provided with balanced sliding sash windows, their earliest known use, glazed with crown glass (Ref. 7.31). Shortly after Wren used the same kind of window at Hampton Court. Sash windows filled with brilliantly clear, often slightly curved, crown glass became a characteristic of Georgian architecture.

Fig. 2. INTERIOR OF A CROWN-GLASS HOUSE.

7.20

The various stages in the process of making crown glass (from William Cooper's *Crown-Glass Cutter and Glazier's Manual*, 1825, reproduced from the ninth edition of the *Encyclopaedia Britannica*, Ref. 4.6, Vol. X, p. 658).

7.9 WATER SUPPLY, SEWAGE DISPOSAL, AND FIRE FIGHTING

In theory the Renaissance reverted to Roman concepts of hygiene. Vitruvius' *Ten Books* (Ref. 2.3) acquired an authority they possibly never had in Roman times. Alberti's *Ten Books,* published as a Latin manuscript in 1452, stressed even more strongly the importance of good water-supply and sewage-disposal systems (Refs. 2.4 and 7.21):

The Ancients had so high a Notion of the Serviceableness of Drains and Sewers, that they bestowed no greater Care and Expence upon any Structure whatsoever, than they did upon them; and among all the wonderful buildings in the City of Rome, the Drains are accounted the Noblest (Ref. 2.4, Book IV, Chapter 7, p. 80).

In practice, however, the effort put by the Renaissance into water supply and sewage disposal was a modest one. Some Roman aqueducts were repaired; for example, in the reign of Pope Sixtus V (1585–1590) Domenico Fontana built the Aqua Felice which was a reconstruction of the ancient Roman Aqua Claudia. More effort was put into new water supply than in medieval times, but this was perhaps due more to the delight of Renaissance princes in spectacular fountains than to a regard for public health.

Bathing was not so rare as many books on the period have suggested. Many Renaissance and Baroque palaces had elaborate bathrooms, and wooden or metal tubs filled with hot water by bucket and emptied in the same way were used elsewhere. At the upper and middle levels of society regular baths were reasonably common. Elizabeth I of England had a bath once a month. Marie Antoinette of France had one every day; her bathroom is still on view at Versailles (Ref. 7.8, pp. 75 and 100).

"Taking the waters" by drinking or immersion became fashionable during the later Renaissance. The "health springs" at Plombières-les-Bains in the Vosges mountains of France were in use in the sixteenth century, and in England the Roman springs of Bath (see Sections 4.3 and 4.6) were rediscovered in the seventeenth. The popularity of the watering resorts produced some fine architecture, particularly at Bath, but the beneficial effect of the baths on public health is open to question.

The main hygienic problem of the Renaissance, as in the Middle Ages, was the failure to recognize the connection between sewage disposal and infectious disease. Muslim medicine had come to western Europe with other scientific texts toward the end of the Middle Ages (see Section 5.4), and the medical schools of the European universities soon added to this knowledge. Although treatment of disease became more expert, avoidance of epidemics made slow progress. As techniques of sewage disposal slowly improved, the cities increased in size and the problems became more difficult.

Paris, then the largest city of Europe, grew from 260,000 in 1553 to 500,000 in 1718 (Ref. 4.6, Vol. XVIII, p. 277). London overtook it in the eighteenth century: 180,000 in 1600 (3.3% of the population of England), 350,000 in 1650 (6.3%), and 550,000 (9.2%) in 1700 (Ref. 4.6, Vol. XIV, p. 821). In 1665 an outbreak of plague is reported to have cost 55,000 lives. Although there was no repetition of the Black Death of the fourteenth century, serious epidemics continued into the nineteenth, when the connection between sewage and certain infectious diseases, such as cholera, was established and enough public money was made available for the construction of proper sewers (see Ref. 1.3, Section 7.5).

The only large sewer built during the Renaissance was in Paris. In the middle of the sixteenth century Henry II asked the French Parliament for money to build a sewer, but Parliament suggested that it should come out of royal revenue and nothing was done. The issue was reopened from time to time, but only in the reign of Louis XIII in the seventeenth century was the money found and Paris became the first European city since Roman times to build a main sewer. London had to wait for its first large sewer until the 1850s. It should perhaps be pointed out that Louis XIII's sewer was a modest effort and that the famous sewer system of Paris also dates from the nineteenth century.

Drains were laid, both in palaces and cities, to discharge sewage and sometimes rain water. The drains, however, commonly discharged into a stream. This kept the sewage off the streets but polluted the water supply.

A connection between disease and bad smells had long been recognized, but a common antidote was to mask the bad odor with a pleasanter one. It was also known that cities posed a public health problem, and in times of epidemics those who could retired to their country houses.

Alberti devoted much more space to hygiene than Vitruvius and also countered some old superstitions:

The Inhabitants of Rome, from the frequent changes of the Air, and the nocturnal Vapours which arise from the River, as also from the Winds which commonly blow very cold about three o'Clock in Summer, at which Time Mens Bodies are extreamly heated, and even contract the very Veins. But in my Opinion these Fevers, and indeed most of the worst Distempers there proceed, in great Measure, from the Water of the Tyber, which is commonly drunk when it is foul (Ref. 2.4, X, 6, p. 218).

On the other hand:

To mend the Air which is unhealthy or corrupted is a Work scarce thought possible to be done by any human Contrivance; unless by appeasing the Wrath of Heaven by Prayers and Supplications (Ref. 2.4, X, 1, p. 211).

An inadequate water supply was also a main cause of the disastrous fires that destroyed large parts of medieval and Renaissance cities at frequent intervals. There were no organized fire brigades until after the Great Fire of London in 1666, which started in a bakery and was first noticed at 2 A.M. on the morning of Sunday, September 2. The Lord Mayor of London, whose duties included the supervision of fire fighting, organized a bucket chain and went back to bed. Small fires were common and were generally extinguished by local people carrying buckets of water. By morning the fire had spread to London Bridge and had destroyed the wooden wheels underneath the bridge which supplied a large part of the city with water (Ref. 3.11, pp. 27–41). The water supply failed and the only defense left was the formation of a vast firebreak by demolishing houses all around the fire. The Lord Mayor was concerned about the possible cost to the city and sought agreement of the owners. When, eventually, he decided on demolition without consent, the fire was completely out of control, and most of the City of London had been destroyed when on September 6 it burned itself out. Macauley, in his *History of England,* described it as a disaster "such as had not been known in Europe since the conflagration of Rome under Nero."

The Great Fire led to more stringent building regulations. Minimum wall thicknesses were specified (see Section 7.8), the use of timber was restricted, and inflammable roof coverings were forbidden. The creation of fire brigades was left to private enterprise.

In 1680 Dr. Nicholas Barbon, who had made his money as a speculative builder, opened a Fire Office and offered to insure property owners against loss by fire. To protect this property he organized a fire brigade manned by fulltime fire fighters who appeared promptly if the house of one of his subscribers caught fire. They would not, however, put out fires in other houses.

The scheme was so successful that by 1698 there were three rival fire insurance companies, each with its own fire brigade, equipped with distinctive uniforms. The practice spread to the Continent and to America, and private fire brigades were continued into the nineteenth century, at which time they were absorbed into a public service.

7.10   OTHER ENVIRONMENTAL ASPECTS

There was some improvement in heating compared with medieval times (see Section 5.9). Fireplaces built during the Renaissance frequently had chimneys that removed the smoke and fireplaces in bedrooms became more common. In the seventeenth century stoves began to appear in Scandinavia, the Netherlands, and Germany (Fig. 7.21). The Roman hypocaust was not revived, however, and comparable modern methods of heating were not invented until the eighteenth century (Section 8.9).

The greater security of the Renaissance (see Section 5.5) made it possible to have larger windows. Because the size of glass was restricted by the manufacturing process (see Section 7.8), windows were assembled from a number of small panes. The regional variations in Renaissance buildings show a sensible appreciation of climate. In Italy window sizes were restricted by the hot summer weather; in England they were enlarged to admit more light in winter (Fig. 7.22).

There were some significant innovations in the seventeenth and eighteenth centuries in the use of natural light as an element of architectural decoration. The study of perspective had been a main preoccupation of the artists of the early Renaissance and had resulted in the production of architectural drawings that showed the three-dimensional appearance of proposed buildings with considerable accuracy. The same accurate perspective is apparent in paintings of the High Renaissance.

This control of perspective was also employed for illusionism. Musical instruments were painted on walls with such accuracy that one walked up to them to pick them up and noticed the illusion only a few feet away. Similarly complete rooms extended the interior where there was only a blank wall and vaults were painted on flat ceilings. The effectiveness of the illusion frequently depended on the careful control of indirect natural lighting.

Perhaps the most extreme example is a Baroque dormer window, added in the early eighteenth century to the ambulatory of the thirteenth-century Gothic cathedral

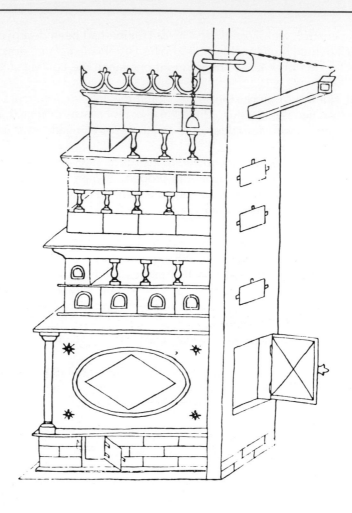

7.21

Swedish tile stove from the seventeenth century (Ref. 7.29, p. 122).

at Toledo (Ref. 7.22, Fig. 79, p. 188), which introduced a bright light into the otherwise gloomy cathedral; in it there appeared a scene of Christ seated on clouds surrounded by a heavenly host, which, although painted on the walls, has a startling three-dimensionality. It is also a tribute to the structural soundness of the Gothic ribbed roof that the eighteenth-century architect Narciso Tome was able to open it up and fit a complete domed room above it without damage to the masonry.

Artificial lighting remained primitive during the Renaissance because of a shortage of suitable fuels for oil lamps and candle making. Materials did not become more plentiful until the eighteenth century.

The theater made a comeback in the Renaissance, although the audiences were smaller. In Vicenza the Teatro Olimpico was designed by Andrea Palladio for the

7.22a

The Farnese Palace, Rome, has small windows which restrict the admission of solar radiation in summer. It dates from the sixteenth century.

Olympian Club, which consisted of sixty-three elected members. It followed in general the arrangement of a Roman theater. The seats are steeply banked, the lowest above the level of the stage, in a semiellipse, not a semicircle (see Section 4.4). The stage therefore can be seen from every seat, for it is quite close to all of them, and the acoustics are good, regardless of the shape of the roof. Palladio designed the theater without a roof, in Roman fashion. A canvas awning was added later, perhaps to preserve the structure, which is built entirely of wood. The present flat roof dates from the nineteenth century. The stage is long and narrow, also in Roman style, and backed by a proscenium designed to show a classical architectural facade. Like the Roman proscenium, it is pierced by five doorways from which radiate passages built as city streets in perspective, so that the streets narrow and the floor rises as one recedes from the stage front. Every spectator can see at least one of these passages, and actors reciting lines in these passages can be heard if their voices carry well enough.

The Olimpico is the only theater surviving from the sixteenth century. In 1618 Giovanni Battista Aleotti designed the Farnese in Parma, which also survives. Here the Roman proscenium remained only as a decoration, and the action took place

7.22b

Longleat House, Wiltshire, also from the sixteenth century, has much larger windows to admit more daylight in winter.

behind a large opening that could be closed by a curtain and changeable scenery painted on movable screens could be introduced.

Instead of using a semicircular or semielliptical plan, as in all Greek, Roman, and Renaissance theaters, Aleotti's auditorium is U-shaped, probably following the arrangement of seating in the masques and court plays staged in banquet halls during the Renaissance. By the end of the century the U had become a horseshoe; the seats lining it were arranged in boxes rising to the ceiling and the floor within it was filled with them. The Imperial Theater of Vienna, designed about 1690 by Ludovico Burnacini in this style (Ref. 7.9, p. 132), remained the standard layout for European auditoria until the twentieth century. There is no indication that any rules, empirical or otherwise, were used for the acoustical design of these auditoria; however, the good sightlines and relative smallness of Renaissance and Baroque theaters would have ensured good audibility.

Music before the nineteenth century tended to conform to the acoustical qualities of the auditoria rather than vice versa. The Gregorian chant sounds best in an auditorium with a long reverberation period, such as a medieval cathedral (see Section 5.9), that may be as high as 10 seconds.

The Renaissance

The Reformation took over the old Catholic churches, many of which had long reverberation periods, but in the newly built Lutheran churches the congregation occupied galleries as well as the main floor, and this may have affected the music of Protestant Germany. Leo L. Beranek (Ref. 7.23, p. 46) estimated the reverberation time of the Thomaskirche in Leipzig, for whose congregation Johann Sebastian Bach composed most of his religious music in the seventeenth century, as 1.6 to 2 seconds.

The fifteenth century did not produce so marked a change in the style of music as the Renaissance did in architecture, painting, and sculpture. The change occurred in the seventeenth century with composers such as Bach and Handel in Germany and Vivaldi and Corelli in Italy. This is the period of Baroque architecture in Italy and Germany. In its own time Baroque secular music was normally performed in small rooms with hard reflecting surfaces, such as ballrooms, small theaters, or large living rooms. Beranek estimated that the reverberation period of these rooms would have been below 1.5 seconds if the room were filled with people. The music therefore had high definition and, because of the many nearby sound-reflecting surfaces, it sounded intimate. Again, it may be argued that the composer adapted himself to the existing architecture and not vice versa.

There was no science of architectural acoustics before the nineteenth century; but its prehistory goes back to the Baroque period. In 1640 Marin Mersenne measured the velocity of sound as 1037 ft/sec (316 m/sec), an error of less than 10%. Galileo discussed the laws of vibration in *Due Nuove Scienze,* published in 1638 (Ref. 7.19). Otto von Guericke demonstrated in 1672 that sound, unlike light, cannot travel in a vacuum and that sound is therefore airborne (Ref. 2.6). In 1650 Athanasius Kircher, better known as the inventor of the "magic lantern" (the precursor of the modern slide projector), discussed the problem of the sound mirror (Fig. 7.23), which is used to explain the whispering galleries found in many Renaissance buildings with circular plans.

Wallace C. Sabine considered that all were accidental:

It is probable that all existing whispering galleries, it is certain that the six more famous ones, are accidents; it is equally certain that all could have been predetermined without difficulty, and like most accidents could have been improved upon (Ref. 7.24, p. 255).

He then analyzed the six in some detail. The most successful, the gallery of the dome of St. Paul's Cathedral in London, forms a circular sound reflector (Ref. 4.7, p. 23), and many other whispering galleries are associated with circular domes.

Sir George Airy, Astronomer Royal in the nineteenth century, ascribed the effect of St. Paul's to the reflection from the surface of the dome overhead, but Lord Rayleigh, a late nineteenth-century physicist, established that the reflection came from the circular wall of the gallery and that the vertical spreading of the sound was restricted by its floor and by the overhanging ledge of the cornice molding. "The whisper seems to creep around the gallery horizontally, not necessarily along the shorter arc, but rather along that arc towards which the whisperer faces" (Ref. 7.24, p. 272).

On the other hand, Sabine noted that not all whispering galleries result from the design of the building; the main sound mirrors in the *Salle des Cariatides* in the Louvre were two vases which were exhibits and not part of the fabric of the building; the horizontal bowls of these vases, in conjunction with the vaulted ceiling, provided the reflectors.

7.23

Focusing of sound in an ellipsoidal chamber, from Kircher's *Musurgia Universalis* (Ref. 7.30, Fig. 2).

It is appropriate to give brief consideration to the theory of music, which derived from classical Greece (see Section 2.2) and was transmitted to the Middle Ages in *De Musica* by Boethius, a sixth-century philosopher. It had a marked effect on the *scientia*, or theory of design, of the Renaissance but none on architectural acoustics.

7.11 PROPORTIONAL RULES

Chapter 5 of Book IX of Alberti's *Ten Books,* entitled "That the Beauty of all edifices arises principally from three Things, namely the Numbers, Figure, and Collocation of the several Members," includes the harmonic rules of proportion:

And indeed I am every Day more convinced of the Truth of Pythagoras's Saying, that Nature is sure to act consistently, and with a constant Analogy in all her Operations: From whence I conclude, that the same Numbers, by means of which the Agreement of Sounds affects our Ears with Delights, are the very same which please our Eyes and our Mind. We shall therefore borrow all our Rules for the finishing our Proportions, from the Musicians, who are the greatest Masters of this Sort of Numbers, and from those particular Things wherein Nature shews herself most excellent and compleat (Ref. 2.4, pp. 197–198).

He then derived a number of ratios from the "Harmony of the Ancients."

In his Italian translation and commentary on Vitruvius, published in Venice in 1567, Daniele Barbaro drew attention to the rules of proportion implicit in Vitruvius' writings. In Book IV, Chapter 4, Vitruvius stated that a temple should be made twice as long as it is wide (which is the octave) and that this double length should be divided between the *cella* and the *pronaos* in the ratio 5:3. In Book VI, Chapter 3, he gave three separate rules for the plan of the atrium of a house. One is again the ratio 5:3, the second is 3:2, and the third is the ratio of the side of a square to its diagonal ($\sqrt{2}:1$).

Palladio (Ref. 3.26, Book I, Chapter 21, p. 27) also gave the Vitruvian ratios 2:1, 5:3, 3:2, $\sqrt{2}:1$, to which he added 1:1 and 4:3, as beautiful proportions. All are harmonic proportions and, except for $\sqrt{2}:1$, commensurable numbers.

The ratio $\sqrt{2}:1$ enters inevitably into the design of an octagonal plan based on the expansion of an eight-pointed star, a method favored by Leonardo da Vinci and used in many designs in his notebooks (Fig. 7.24).

The rules of proportion have been discussed in detail by Schofield (Ref. 7.25) and Wittkower (Ref. 7.26); the latter has shown that they can be found in the dimensioned drawings of Palladio's buildings, reproduced in the second of his *Four Books* (Ref. 3.26), and also in Alberti's surviving buildings.

There is no *a priori* reason why proportions that delight the ears should also delight the eyes, and there is no evidence that buildings produced by harmonic rules, or by any other proportional rules, are more beautiful on that account. The proportional rules of the Renaissance, however, reprinted many times (e.g., in the first English book on architecture by Wotton, Ref. 2.5), provided a system of reference for design and a method of modular coordination. The other system of reference was the proportional rules of "the Orders," which formed the central feature of most books on architecture from the fifteenth to the seventeenth centuries and were derived from Vitruvius' Book III (see Sections 2.1 and 2.2).

The Renaissance also produced the first written rules for structural design, although we noted in Sections 6.3 and 6.7 that these rules may have existed earlier.

Alberti described rules for sizing piles and arches:

These piles should never be less in length than the eighth part of the Height of the Wall to be built upon them, and for their Thickness, it should be the twelfth Part of their Length, and no less (Ref. 2.4, Book II, Chapter 3, p. 44).

There should not be a single Stone in the Arch but what is in Thickness at least one tenth part of the Chord of that Arch; nor should the Chord itself be longer than six Times the Thickness of the Pier, nor shorter than four Times. The Stones also should be strongly fastened together with Pins and Cramps of Brass. And the last Wedge, which is called the Key-stone, should be cut according to the Lines of the other Wedges, but left a small Matter bigger at the Top, so that it may not be got into its Place without some Stroke of a light Beetle; which will drive the lower Wedges closer together, and so keep them tight to their Duty (Ref. 2.4, Book IV, Chapter 6, pp. 78–79).

The ratio span/thickness = 10 given by Alberti is identical to that calculated by Heyman for the semicircular masonry arch based on limit design (Ref. 5.16, p. 232). This could be due to good observation of failures by Alberti over an extended period or it could be a coincidence.

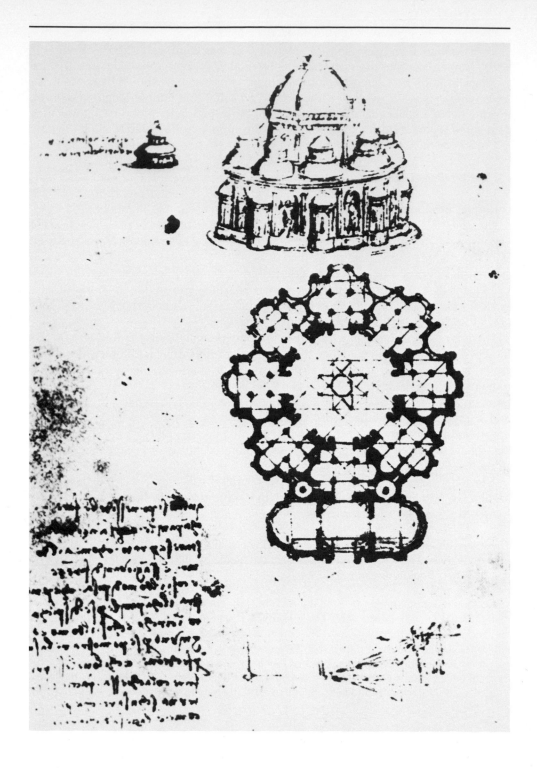

7.24

Design for an octagonal church, based on the expansion of an eight-pointed star and thus involving the ratio $\sqrt{2}:1$. It is one of many drawn by Leonardo da Vinci (from MS 2037 in the Bibliothèque Nationale, Paris, folio 5, verso).

Another rule for determining the thickness of a pier was given by François Blondel in *Cours d'architecture,* published in Paris between 1675 and 1683 (Ref. 7.27, p. 92). A trapezium with three equal sides *a* is inscribed in the arch and on one side is extended downward by its own length *a*. The thickness of the pier is then made equal to the vertical projection *b* of the length *a* (Fig. 7.25). For a semicircular arch the answer is 4.1; Alberti gave "between 4 and 6." Blondel's rule was commended by B. F. de Bélidor in *Science des ingénieurs,* published in Paris in 1729, a mathematical textbook on the theory of structures still in use in the nineteenth century, and was employed by Viollet-le-Duc in his restoration of Gothic structures (see Section 6.5). Blondel was also a noted proponent of harmonic proportions (Ref. 7.25, p. 70).

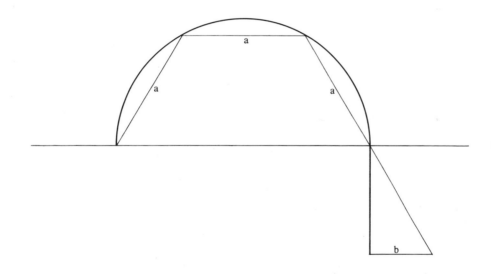

7.25

Blondel's rule for determining the thickness of the abutment of a vault. Inscribe a trapezium with three equal sides *a* inside the arch and extend one side downward by *a*. The thickness of the pier is the vertical projection *b* of this line *a*.

Practical geometry developed rapidly in the seventeenth century. The geometry of Vitruvius and Villard had been relatively simple (see Sections 2.2 and 6.3), but as the Renaissance changed to Mannerism and Baroque the geometric constructions became more and more complex (Fig. 7.26).

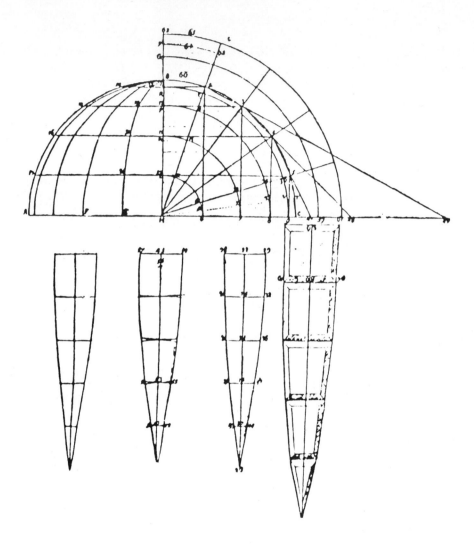

7.26

Construction for developing the surface of a dome so that the roof sheeting can be cut (from Guarini's *Architettura Civile* (Ref. 2.13, Fig. 3, Table 41)).

CHAPTER EIGHT

The Age of Reason

and the

Industrial Revolution

Knowledge is of two kinds.
We know the subject ourselves,
or we know where we can find
information upon it.

SAMUEL JOHNSON
in 1775

Boswell's Life of Johnson

Science and technology made great advances in the eighteenth and nineteenth centuries. During the period reviewed in this chapter (1700 to 1815) the foundations were laid for the mathematical theory of structural design, the first iron structures were built, and concrete was rediscovered.

The factory buildings erected at the end of the eighteenth and in the early nineteenth centuries are the prototypes of modern architecture. Entirely new building types, such as hospitals with separate wards and mass-produced houses, appeared during that time. The churches and palaces of the period, however, produced few technical innovations.

8.1 THE AGE OF REASON AND THE FRENCH REVOLUTION

In the early sixteenth century it was still possible for a well-educated man to be conversant with all the scientific and technical books published in western Europe, but science made rapid progress during the seventeenth century and in the eighteenth encyclopaedias were produced to systematize the new knowledge. The best known of these were Ephraim Chambers' *Cyclopaedia,* first published in London in 1728 in two folio volumes, Johann Heinrich Zedler's *Grosses Vollständiges Universal Lexicon,* published in Leipzig between 1732 and 1750 in sixty-four volumes, and the *Encyclopédie* edited by Denis Diderot and Jean le Rond D'Alembert in Paris (1751–1772) in twenty-eight volumes. Diderot and D'Alembert, unlike previous encyclopaedists, included "the trades" and "the mechanical arts" and named their contributors, among whom were some of the best French experts. The authors of the relevant articles were expected to visit workshops and not merely rely on other books, as Chambers had done. The *Encyclopédie* met with much opposition from the church and the government of Louis XV because of its philosophical entries; an early objection was made to the article on *certitude* which was based on Locke's analysis of knowledge and rejected biblical revelation.

The willingness of the eighteenth-century philosophers and scientists to look at all problems from first principles produced great changes in the various branches of science, including structural mechanics; it was also a major cause of the French Revolution in 1789.

During the seventeenth century France became the leading military power in western Europe. Sébastien Le Prestre de Vauban, best known for his design of fortifications, was appointed *ingénieur de roi* in 1655 and *maréchal de France* in 1703; at his suggestion the Minister of War in 1675 created the *Corps des ingénieurs du génie militaire,* which was also employed on a number of important civilian projects such as road construction and the water supply for Versailles (see Section 8.3). The *Corps des ingénieurs des ponts et chaussées* (corps of bridge and road engineers) was established in 1716, and the first technical university, the *École des ponts et chaussées,* was founded in 1747. Jean Perronet became Inspector General of the Corps in 1750 and as such was responsible for the supervision of the *École;* he was a distinguished theoretician, as well as a practical engineer who built about thirty bridges, including the Pont de Neuilly and the Pont de la Concorde in Paris. In 1760 he reorganized the system of education; students were required to learn mathematics before being allowed to study engineering.

The French Revolution abolished the schools and universities of the Old Regime, and in 1795 several engineering schools were combined to form the *École Polytechnique,* headed by the mathematician Gaspard Monge as its first director. The new school emphasized even more strongly the scientific basis of engineering. Two years were devoted to mathematics, physics, and chemistry. The Revolution and the Empire had a great need for military engineers and the engineering schools prospered. The most theoretically based technical education in the world, acquired first by France, was copied in the nineteenth century by the rest of Europe.

Already during the sixteenth century the leading scientists had corresponded with one another. The first scientific societies were founded in the seventeenth: the *Academia dei Lincei* in Rome (1600–1630; the present academy of the same name was founded in 1870) and the equally short-lived *Academia del Cimento* in Florence (1651–1667); the *Royal Society of London* (1666) and the *Académie des Sciences* in Paris (1667) are still flourishing.

The English and French societies continued to exchange information during the Revolutionary and Napoleonic wars. English and Continental scientists continued their correspondence and even exchanged occasional visits; for example, Sir Humphry Davy was awarded Napoleon's prize by the *Institut de France* in 1806, even though France and Britain were at war, for his work on galvanic electricity. In 1813, when the war was at its height, he visited Paris with his wife at Napoleon's invitation and with the consent of the British government to collect his medal. He took with him a portable laboratory and in December 1813 set it up in a factory for saltpeter (a principal ingredient of gunpowder) whose owner Cortois had discovered a new substance now called iodine. Science was not yet part of the war effort.

*8.2   THE STRENGTH OF MATERIALS

Testing the strength of materials started tentatively in the early Renaissance (see Section 7.4) but was not undertaken systematically until the seventeenth century. Tests were done mostly on full-size specimens, frequently by the use of cannon balls, the only heavy and accurate weights available in quantity (Fig. 8.1).

In the eighteenth century testing machines that employed levers to multiply the effect of a small weight (Fig. 8.2) were devised. In addition, specimens were made as small as possible.

Such tests were conducted by a number of physicists who tabulated the tensile, compressive, and flexural strength of various types of timber, metal, glass, and other material within the capacity of the machines. The best-known tables were those published by Petrus van Musschenbroek, professor of physics at the University of Leyden, in *Physicae experimentales et geometricae dissertationes* (Leyden, 1729). This book contains a long chapter, *Introductio ad cohaerentiam corporum firmorum* (Introduction to the Cohesion of Solid Bodies), summarized by Todhunter (Ref. 8.15, Vol. I, p. 16). Musschenbroek's data were held in high regard throughout the eighteenth and early nineteenth centuries and were used, for example, in the Le Seur-Jacquier-Boscovich report on the safety of the dome of S. Pietro (see Section

---

* Sections 8.2, 8.3, and 8.4 assume some knowledge of mechanics. They can be omitted without loss of continuity of the story.

8.1

Testing the strength of a full-size timber beam, using cannon balls as weights (from *An Essay in the Strength and Stress of Timber*, by Peter Barlow. Printed for A. J. Taylor at the Architectural Library, London 1817, Plate V).

7.6; Poleni in a similar investigation made his own tests on iron). Musschenbroek's tables for the strength of timber were still reprinted in the Sixth Edition of *A Treatise on the Strength of Materials* by Peter Barlow F. R. S., professor at the Royal Military Academy, published by Lockwood in London in 1867.

Sir Isaac Newton, in the second edition of *Optics or a Treatise of the Reflections, Refractions and Colours of Light* (published in 1717), posed a query on the strength of materials:

The parts of all homogeneal hard bodies, which fully touch one another, stick together very strongly. And for explaining how this may be, some have invented hooked atoms, which is begging the question; and others tell us, that bodies are glued together by Rest: that is, by an occult quality, or rather by nothing: and others, that they stick together by conspiring motions, that is by relative rest among themselves. I had rather infer from their cohesion, that their particles attract one another by some force, which in immediate contact is exceedingly strong, at small distances performs the chemical operations above-mentioned, and reaches not far from the particles with any sensible effect (Ref. 8.15, pp. 13–14).

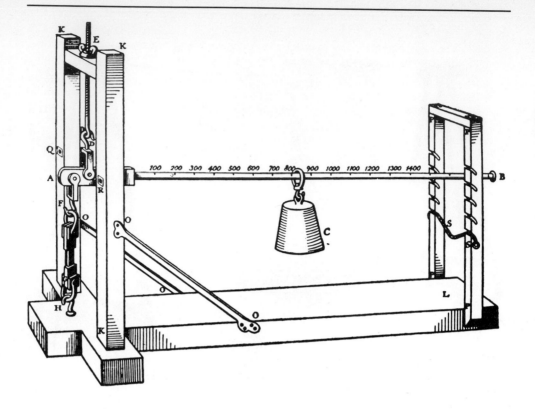

8.2

Machine for testing small tension specimens. The load was multiplied by a lever. The test specimen was gripped in shackles which were self-aligning to ensure concentric loading (from *Physicae experimentales et geometricae dissertationes,* by Petrus van Musschenbroek, Leyden 1729).

Musschenbroek argued in his abovementioned book that it was not necessary to have a physical or metaphysical hypothesis for the nature of the *vires internae* (internal resistance forces) in order to use them in a practical problem. They could be determined satisfactorily by a testing machine. He also held that elasticity, discussed in Section 8.3, was a *vis interna attrahens* (an internal attracting force). Both concepts are still valid today.

Musschenbroek treated separately the tensile strength from the direct tension test *(Cohaerentia vel resistentia absoluta)* and the bending test *(Cohaerentia respectiva aut transversa);* there was still no satisfactory theory of flexure (see Section 8.3).

The small size of Musschenbroek's timber specimen was criticized by the Comte de Buffon, director of the Royal Botanical Gardens in Paris and a contributor to the *Encyclopédie,* who found considerable variation in the strength of wood specimens taken from different parts of the same tree and suggested that only large specimens

　　　　　　　　The Age of Reason and the Industrial Revolution

would give reliable information. We have, however, accepted Musschenbroek's approach with a proviso that an adequate number of tests be made; Leonardo had already tentatively expressed the same view (see Section 7.4).

Special tests were made on the strength of the building stones to be used in the construction of the Church of St. Geneviève by the engineer Emiland Marie Gauthey (a student of Perronet, Section 8.1) in conjunction with the architect Soufflot. This was probably the first time that this had been done for a building.

The dome of St. Geneviève, renamed the Panthéon when the Revolution secularized the churches, is not large compared with the three discussed in Chapter 7. Its diameter is 21 m (69 ft). As in the slightly larger dome of the Hôtel des Invalides, built in the preceding century, there is an inner dome through which one can see a middle dome; above there is a third dome to be seen from the outside. The triple dome was designed by traditional methods, but during the prolonged construction (1755–1781) a number of criticisms were made about the strength of the piers, which are slender by comparison with those of earlier domes. Gauthey calculated the thrust of the dome, using de La Hire's method (Section 7.6) and found that the compressive strength of the stone provided a factor of safety of at least 10.

Another novel feature of this structure was the systematic use of wrought iron bars embedded in the masonry as *armature* (Fig. 8.3). Soufflot may thus be regarded as one of the pioneers of reinforcement in masonry structures.

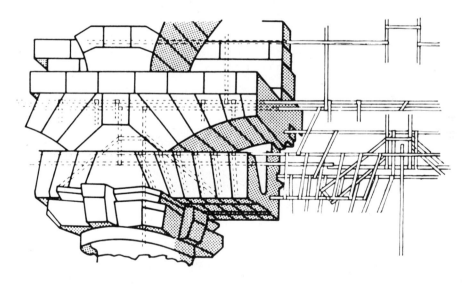

8.3

Reinforcement in the masonry of St. Geneviève, Paris, about 1770.

8.3 THE ELASTICITY OF MATERIALS AND THE THEORY OF BENDING

The discovery of elasticity is credited to Robert Hooke, better known for his work on microscopy; he was also a City Surveyor and the architect of some of the churches built after the Great Fire (Section 7.3). In his book on helioscopes (Section 7.3) he printed among his anagrams:

3. The True Theory of Elasticity or Springiness and a particular Explication thereof in several subjects in which it is to be found. And the way of computing the velocity of bodies moved by them. ceiiinosssttuu.

This he later interpreted as *Ut Tensio sic vis* (as the extension, so is the force); today we would say "strain is proportional to stress." The object of this anagram apparently was to protect his rights to the observation in case his work on springs should lead to a patent for use in watches.

In 1678 he published his findings in one of the Cutlerian Lectures delivered before the Royal Society of London. This was printed in the same year under the title *De Potentia Restitutiva, or of Spring Explaining the Powers of Springing Bodies*. He illustrated the experiment (Fig. 8.4) and described it as follows:

Take a wire string of 20, or 30, or 40 feet long, and fasten the upper part thereof to a nail, and the other end fasten a Scale to receive the weights: Then with a pair of compasses take the distance of the bottom of the scale from the ground or floor underneath, and set down the said distance, then put in weights into the said scale and measure the several stretchings of the said string, and set them down.

Then compare the several stretchings of the said string, and you will find that they will always bear the same proportions one to the other that the weights do that made them.

This led him to the following conclusions:

It is very evident that the Rule or Law of Nature in every springing body is, that the force or power thereof to restore itself to its natural position is always proportionate to the Distance or space it is removed therefrom, whether it be by rarefaction, or separation of its parts the one from the other, or by a Condensation, or crowding of those parts nearer together. Nor is it observable in these bodies only, but in all other springy bodies whatsoever, whether Metal, Wood, Stones, baked Earths, Hair, Horns, Silk, Bones, Sinews, Glass and the like (Ref. 8.16, p. 19–20).

One of Hooke's examples of a springy body in this paper was a piece of dry wood bent in such a way that the top of the beam was extended and the bottom compressed (Fig. 8.5). The originally plane and parallel sections remain plane and converge on a common center of curvature. This is, in fact, Navier's hypothesis of 1826, discussed later in this section. Hooke was an accurate experimenter but he did not follow his observation through to correct Galileo's theory of bending.

Hooke's law made two distinct statements: one about the recovery of the elastic deformation of structural materials and the other about the linear relation between the applied load and the elastic deformation. In the early nineteenth century these

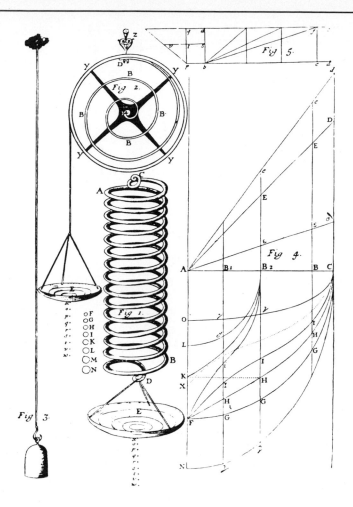

8.4

Robert Hooke's illustration of his tests on springs and elastic wires (from *Of Spring*, published by John Martyn, London, 1678).

statements became the basis of the classical theory of structures. The first success was in Euler's theory of buckling, which we discuss in Section 8.4, but the first attempt to use them was in relation to the important problem of bending, already considered by Leonardo and Galileo (Section 7.4).

Galileo's formula for the strength of a beam is

$$M = \tfrac{1}{2}fbd^2$$

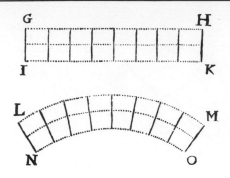

8.5

Another illustration from Robert Hooke's *Of Spring* shows a piece of dry wood bent so that the top of the beam is extended and the bottom compressed. The originally plane and parallel sections remain plane and converge on a common center of curvature. This assumption was used subsequently by L. M. H. Navier as the geometric basis for his theorem of bending. He may or may not have derived it from Hooke. Hooke used the figure to illustrate springiness, not to solve the theory of bending.

where M is the bending moment (i.e., the moment produced by the load or loads carried by the beam), f is the stress, and b and d are the width and depth of the rectangular section (Fig. 7.9).

This was challenged by Edmé Mariotte, codiscoverer of the gas law which in France bears his name but in England is known as Boyle's law. Mariotte had been charged with the design of the water-supply pipes for the Palace of Versailles and had made an investigation of the thickness of the pipes required. As a result he tested pipes both in tension and bending and, using Galileo's formula, found that the strength of the material obtained from tension tests was much higher than that obtained from bending tests.

He pointed out that Galileo had assumed implicitly (see Section 7.4) that his cantilever was an inextensible solid. Mariotte argued that even the hardest materials deform under load and assumed, with Hooke, that they were elastic and that load was proportional to deformation. He then observed that there must be some compression at the bottom of the cantilever. The beam might thus be considered as a bundle of individual fibers that deformed in tension near the top of the cantilever and in compression near the bottom, the extension or compression depending on the load. These assumptions are all basically correct.

Mariotte then made a mistake in his arithmetic and got a factor of $\frac{1}{3}$ where Galileo had obtained a factor of $\frac{1}{2}$; the correct answer is $\frac{1}{6}$ [see Fig. 8.6, Eqn. (8.11)]. The results obtained from his tests agreed with the incorrect formula which is presumably the explanation of the error.

Like all his contemporaries, Mariotte was thinking in terms of ultimate strength. He had conducted a test and was deriving a formula to predict the load at which the test specimen would fail. He assumed that the material would behave elastically, with its load proportional to its deformation, right up to the point of failure. At a certain deformation it would fail. The object was to ensure that the structure would *not* fail, but for this he needed the failing load. Barring the arithmetic error, this was a correct elastic theory, but his supposition that it would agree with ultimate strength tests is wrong. A few brittle materials behave elastically up to fracture in tension, but they have a higher compressive strength so that the neutral axis moves toward the compression face during the test. All other materials lose their elasticity long before failure and the relation between load and deformation then ceases to be linear. In Mariotte's time it was not possible to derive an ultimate strength theory for bending—only an elastic theory. On the other hand, it was possible only to test for ultimate strength; an elastic theory could not be checked. Mariotte published his findings in *Traité du mouvement des eaux* (Treatise on the Movement of Water) in 1686.

The history of the solution of the theory of bending is a long and tortuous one and has been recounted in detail by Todhunter (Ref. 8.15), Heyman (Ref. 7.16), more briefly by Timoshenko (Ref. 8.16), and by Straub (Ref. 3.36). A number of people came close to solving the problem.

Mariotte in his concept of fibers had, in fact, formulated the concept of stress as a force acting on a very small area. A. Parent, an *élève* (assistant) at the French Academy, published in 1713 three volumes of essays containing partly *memoires* previously published by the *Académie* and partly rejected papers in which he applied the infinitesimal calculus to the fibers and thus obtained the section modulus for the circular section. He noted that the resultant force due to the fibers in tension must equal the resultant force due to the fibers in compression and corrected Mariotte's error; that is he noted that the factor [in Fig. 8.6, Eqn. (8.10)] is $\frac{1}{6}$.

He then observed that the neutral axis may, for some materials, move close to the compression face *during* the fracture of the beam, but it will not be found there *before* fracture; indeed he appears to have been the first to distinguish between elastic theory and ultimate strength. To get better agreement between the elastic theory of bending and the existing ultimate strength data he assumed different elastic properties for tension and compression. His conclusion therefore was a factor that differed from $\frac{1}{6}$.

Coulomb (Ref. 7.16), whose work we have already considered (Section 7.6) also obtained the factor of $\frac{1}{6}$ (8.11), apparently without knowing Parent's work. He accepted it as suitable for timber but noted that it did not fit the experiments on stone.

But if it is assumed that the member, when about to break, is composed of stiff fibres, that can be neither compressed nor extended; if it is assumed further that the body fractures by rotation about a point *h*, then each point of the depth *fh* will be subject to the same stress (p. 47).

This was a return to Galileo's theory and he obtained the same result, namely, a factor of $\frac{1}{2}$ (8.10).

In 1807 Thomas Young published *A Course of Lectures on Natural Philosophy and the Mechanical Arts*, which contained the substance of a series of lectures delivered at the Royal Institution, founded in London in 1799, for diffusing the knowledge and facilitating the general and speedy introduction of new and useful mechanical inventions and improvements and also for teaching, by regular courses of philosophical lectures and experiments, the applications of new discoveries in science to the improvements of arts and manufactures and in facilitating the means of procuring the comforts and conveniences of life.

These lectures contained several equations for the bending of beams, including $M = \frac{1}{6} fbd^2$, but without proof. Young probably obtained that equation from Coulomb, whose biography he contributed to a supplement for the *Encyclopaedia Britannica*.

With regard to structural safety, Young noted that (Ref. 8.16, p. 93)

a permanent alteration of form . . . limits the strength of materials with regard to practical purposes, almost as much as fracture, since in general the force which is capable of producing this effect is sufficient, with a small addition to increase it till fracture takes place.

This is the conclusion that Navier reached in France at about the same time, as we note later in this section.

Young was quoted by Tredgold (Section 8.6) and other practical engineers in Britain and thus influenced the design of early iron structures.

In the same book Young defined the modulus of elasticity, the constant in Hooke's law and now frequently called Young's modulus. He determined the modulus of elasticity of steel by measuring the vibrations of a steel tuning fork and obtained 29,000 ksi (200 GPa), which is the value still used today.

Young's definition sounds unnecessarily involved (Ref. 8.15, Vol. I, p. 82). In modern terminology,

$$f = Ee \qquad\qquad (8.1)$$

where f is the stress or force per (infinitesimally small) unit area, E is the modulus of elasticity, and e is the strain or elongation per (infinitesimally short) unit length.

Young was a man of exceptional versatility whose substantial contributions to mechanics were overshadowed by his other achievements. He was qualified as a medical practitioner, understood six ancient and several modern languages, and deciphered part of the Rosetta Stone in the British Museum, which provided the first key to the Egyptian hieroglyphs; he is best remembered for his work on optics, color vision, and color blindness.

The elements for a solution were now available. The derivation of the theory still used today and bearing his name was published in Paris in 1826 by Louis Marie Henri Navier, in *Résumé des leçons données à l'École des Ponts et Chaussées, sur l'application de la mécanique a l'etablissement des constructions et des machines* (Summary of Lectures Given at the School for Bridges and Roads on the Application of Mechanics to the Design of Structures and Machines). Navier began these lectures in 1819, and his book remained the standard work on structural theory for most of the nineteenth century; it appeared in its third edition in 1864, twenty-eight years after his death, with notes by Barré de Saint-Venant, one of the greatest contributors to the theory of elasticity.

Navier assumed that the material was elastic and obeyed Hooke's law, but he conducted his own tests to determine the modulus of elasticity of iron before using his theory for the design of an iron bridge.

He assumed that initially plane sections remain plane and converge on a center of curvature (Fig. 8.5). This assumption could not be proved until the late nineteenth century when sufficiently accurate instruments for measuring strain had been developed (see Ref. 1.3, Section 5.2).

He assumed, like Mariotte, that the individual fibers of the beam could move freely in relation to one another and noted that the condition of horizontal equilibrium determined the location of the neutral axis, as Parent had already done: he showed that it passed through the centroid of the cross section.

Navier's theorem followed from these premises (Fig. 8.6). The theorem is valid only as long as the material behaves elastically. Navier considered that structures should behave elastically under the loads they normally carried so that permanent deformation would not be suffered. The maximum stress should therefore not exceed a stress now called the maximum permissible stress and specified in building codes. To provide a factor of safety the maximum permissible stress is substantially lower than the stress at which the material ceases to behave elastically.

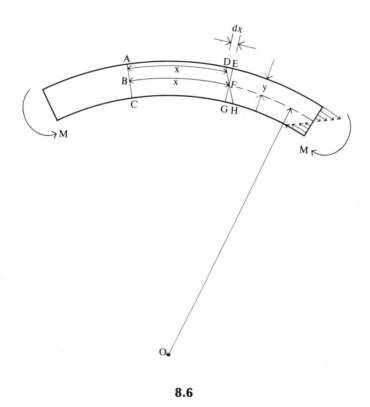

8.6

Navier's theorem of bending.

The beam, under the action of a uniform bending moment M, bends into a circular arc (Fig. 8.5) whose radius of curvature is $R = OF$. Originally plane and parallel sections ABC and EFG thus converge into a common center of curvature O.

This causes tensile strains at the top and compressive strains at the bottom. The line of zero strain BF is called the neutral axis. Navier showed that it passes through the centroid of the section; thus for a symmetrical section it is at half depth.

The maximum tensile strain at a distance y above the neutral axis is $DE/AD = dx/x$, which means that the strain $e = dx/x$.

Let us call the distance of the top fiber from the neutral axis $EF = y$. We noted that $OF = R$. Because the triangles DEF and BFO are similar, $DE/BF = EF/FO$, which gives the strain

$$e = \frac{dx}{x} = \frac{y}{R} \tag{8.2}$$

We now determine the stress by using Hooke's law and Young's modulus of elasticity E:

$$f = Ee = \frac{Ey}{R} \tag{8.3}$$

The stress varies proportionately to the distance y from the neutral axis; it is tensile above and compressive below.

The force acting on an infinitesimally small area dA at a distance y from the neutral axis is

$$dP = f \; dA$$

and the moment of this force about the neutral axis is

$$dM = y \; dP = yf \; dA = \frac{Ey^2}{R} \; dA \tag{8.4}$$

The total resistance moment of the section is the integral of all the infinitesimally small elements dM.

$$M = \int \frac{Ey^2}{R} \; dA \tag{8.5}$$

Because Young's modulus is a constant and the radius of curvature does not vary with the depth y, we can take them outside the integral sign:

$$M = \frac{E}{R} \int y^2 \, dA = \frac{E}{R} I \qquad (8.6)$$

where I is a geometric property of the cross section, called its second moment of area.

Substituting from (8.3) for E/R

$$\frac{M}{I} = \frac{f}{y} \qquad (8.7)$$

and

$$\frac{M}{I} = \frac{f}{y} = \frac{E}{R}$$

This is Navier's theorem.

The theorem is frequently written in the form

$$M = fZ \qquad (8.8)$$

where $Z = I/y$ is called the section modulus.

For a rectangular section of width b and depth d

$$I = \int_{-\frac{1}{2}d}^{+\frac{1}{2}d} by^2 \, dy = \frac{bd^3}{12} \qquad (8.9)$$

and the section modulus

$$Z = \frac{I}{\frac{1}{2}d} = \tfrac{1}{6}bd^2 \qquad (8.10)$$

Thus

$$M = \tfrac{1}{6} fbd^2 \qquad (8.11)$$

which is $\tfrac{1}{3}$ of the value obtained by Galileo and $\tfrac{1}{2}$ of the value obtained by Mariotte.

For the cantilever carrying a concentrated load at its end (Galileo's problem)

$$M = WL = \tfrac{1}{6} fbd^2 \qquad (8.12)$$

The search for a formula to calculate the ultimate strength of a beam, which had occupied many in the seventeenth and eighteenth centuries, was ended by Navier's book. The elastic theory of structures developed during the remainder of the nineteenth and early twentieth centuries to provide a quantitative method for the design of structures that could be used without constant resort to full-scale testing. It completely transformed engineering practice and later architectural design. It reached its greatest prestige during the 1920s and 1930s when it seemed to provide a system for structural design capable of unlimited further perfection. It was only as a result of our experience with structural damage during World War II and subsequent research that we developed qualms about the adequacy of elasticity as the sole basis of structural design and recognized in the confusion of the pioneers of the Age of Reason a basic structural problem that requires further consideration (see Ref. 1.3, Section 4.6).

8.4   THE BUCKLING PROBLEM AND THE THEORY OF ELASTICITY

Euler's buckling formula, published in 1757, is the oldest structural formula still in use today, yet its derivation is quite difficult. Teachers know that it must be presented with some care to a class of engineering students and that a full mathematical derivation is unsuitable for an architectural class.

The formula throws an interesting light on the degree of sophistication reached by applied mathematics by the mideighteenth century and also on the philosophy of the people who used it.

Leonard Euler was born near Basel, in German-speaking Switzerland, in 1707. In 1733 he went to St. Petersburg (now Leningrad) as head of the department of mathematics of the Russian Academy of Science. In 1741 he accepted an invitation from Frederick II (who was also a patron of Bach and Voltaire) to move to the Prussian Academy of Science in Berlin.

The original suggestion for the solution came from Daniel Bernoulli, another Basel-born mathematician who wrote to him in St. Petersburg:

Ew. reflectiren ein wenig darauf ob man nicht könne sine vectis die curvaturam immediate ex principiis mechanicis deduciren. Sonsten exprimire ich die vim vivam potentialem laminae elasticae naturaliter rectae et incurvatae durch $\int ds/R^2$, sumendo elementum ds pro constante et indicando radium osculi per R. Da Niemand die methodum isoperimetricorum so weit perfectionniret als Sie, werden Sie dieses problema, quo requiritur ut $\int ds/R^2$ faciat minimum, gar leicht solviren (Ref. 8.15, p. 30).

The following is an approximate translation of this mixture of Latin and German, used throughout the extensive published correspondence between Bernoulli and Euler:

Perhaps you would like to give a little thought to the question whether the actual curvature could be derived by the use of mechanical principles without using moments. As an alternative, I would express the strain energy of an elastic lamina, which in its normal state is straight and not curved, by $\int ds/R^2$, integrating the element ds for a constant and fixed radius of curvature R. Since no one mastered the calculus of variations as well as you, you will be able to solve this problem very easily, since it requires that $\int ds/R^2$ should be made a minimum.

The calculus of variations is concerned with determining the points on a curve at which it reaches a maximum or a minimum. It was largely developed by Euler and widely used in the eighteenth century for solving a variety of problems: for example, in Coulomb's *Essai sur une application des règles de Maximis et Minimis &c* (Ref. 7.16). The principle of least work or least action was then used to find the greatest load a structure could carry or its most effective geometric form.

Euler in his book on elastic curves, justified the principle:

Since the fabric of the universe is most perfect, and is the work of a most wise Creator, nothing whatsoever takes place in the universe in which some relation of maximum and minimum does not appear. Wherefore there is absolutely no doubt that every effect in the Universe can be explained as satisfactorily from final causes, by the aid of the method of maxima and minima, as it can from the effective causes themselves (Ref. 8.16, p. 31).

In the following century Todhunter commented:

On the whole it is probable that physicists have to thank this theological tendency in great part for the discovery of the modern principles of Least Action, of Least Constraint and perhaps of the Conservation of Energy (Ref. 8.15, p. 34, written before 1884).

Euler took up Bernoulli's suggestion and presented his results in 1759 in a paper *Sur la force des colonnes* published in the *Mémoires de l'Académie de Berlin*. He first obtained a general solution for the curve of a loaded elastic lamina and then derived the special solution for the greatest load P that a straight concentrically loaded column can carry before bending sideways (i.e., buckling):

$$P = \frac{C\pi^2}{4L^2} \qquad (8.13)$$

where π is the circular constant (3.141), L is the length of the column, and C is a quantity that Euler called the "absolute elasticity," adding that it must be determined by experiment. Since the time of Navier (Section 8.3) we have put $C = EI$, where E is Young's modulus and I is the second moment of area (Fig. 8.6).

The Euler formula had already been derived experimentally by Musschenbroek in 1729 (Fig. 8.7, see also Section 8.2). It is correct for a very slender column but not for the sort of column normally used in an architectural structure. In the nineteenth century it became the cause of a major controversy between British and French authors on structural design.

Daniel Bernoulli in 1741 had already solved his own equation for bending moments as part of a study of vibration. Euler rewrote this equation in terms of the "absolute elasticity" C in the form

$$C \frac{d^2y}{dx^2} = M \qquad (8.14)$$

where M is the bending moment and y is the deflection produced by the moment at some point x along the beam.

This equation has been used since Euler's time to calculate the deflection of simple beams.

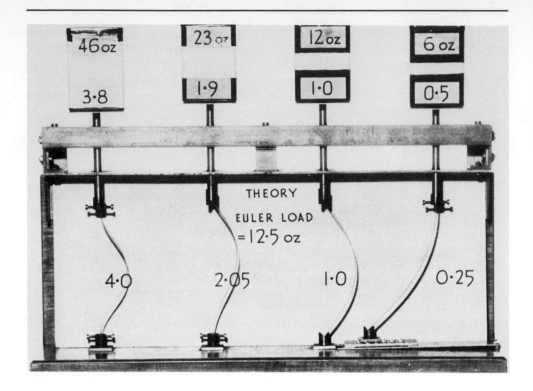

8.7

Buckling of slender columns of the type considered in Euler's theory. A simply supported column (second from the right) had been tested 30 years earlier by Musschenbroek, so that the result was known by experiment. The more the end restraint, the greater the buckling load. Very slender columns buckle by bending sideways and recover their shape elastically, as Euler concluded. No permanent damage results. Columns in practical buildings do not behave that way, however (see 8.17).

In 1826 Navier wrote the equation in terms of EI

$$EI\,\frac{d^2y}{dx^2} = M \tag{8.15}$$

He used it for a variety of more complicated deflection problems and also for the first solution of statically indeterminate structures (Ref. 1.3, Section 3.2).

As we noted in Section 8.3, Navier put structural design firmly on an elastic basis. In the process he shed completely new light on some old problems; for example, he solved the problem of the arch in terms of elasticity. Using (8.15), he obtained the displacement of an arch and then determined the horizontal thrust necessary to prevent it if the arch were hinged at its two supports. From this he calculated both the thrust and the bending moment. The solution cannot be applied to masonry arches, which do not behave elastically at any time because of the joints between the masonry blocks, but it is still used for steel and reinforced concrete arches.

By the end of the nineteenth century it was possible to design most statically determinate structures* as well as some simple statically indeterminate structures.* These design methods are discussed in detail in Ref. 1.3, Chapters 2 and 3.

8.5   THE INDUSTRIAL REVOLUTION

Many historians consider that the French Revolution began on May 5, 1789, when the Estates General (the French parliament) met at Versailles. All agree that it started in 1789. The people who lived through it appreciated that they had witnessed an upheaval.

No firm date can be set for the beginning of the Industrial Revolution, and it is doubtful that the people who witnessed it considered the changes revolutionary. The term was used by a number of economists and historians, of whom Arnold Toynbee is the best known, in the late nineteenth century and gained general acceptance in the twentieth.

The effect of the Industrial Revolution on architecture was much greater than that of the French Revolution. In the first place it produced the iron frame. This was the essential counterpart to the theory of structural design which was being developed during the eighteenth and nineteenth centuries, mainly in France. Within the brief space of a century and a half these two developments completely transformed the design and appearance of buildings. If we compared the best architecture of the second century A.D., the eighteenth century, and the twentieth century, we would find far less difference between the first two than between the last.

The second effect was even more important. The Industrial Revolution increased the population and concentrated it in cities (Table 8.1).

This rapid growth, which began in the mideighteenth and continued throughout the nineteenth century, was accompanied by an equally rapid urbanization. In 1550 the population was still 80% rural, but in the mideighteenth the process of urbanization accelerated. Fifty per cent of the population of England was urban by 1850, 75% by 1900, and 80% by 1939. The proportion has remained steady since then.

In 1750 only two English cities had a population of more than 60,000: London with 750,000 and Bristol with 60,000, or one-eighth of the population of England. In 1870 there were eighteen cities with a population of more than 60,000, or one-third of the population of England. London now had 2.8 million, followed by Liverpool with 440,000, and Manchester and Salford with a combined population of 440,000.

The twin cities of Manchester and Salford grew particularly fast during the Industrial Revolution (Ref. 4.6) (Table 8.2). This produced a great need for new buildings

* See Glossary.

Table 8.1

Year	Population of England
1400	
(before the Black Death)	Approx. 3.5 million
1550	Approx. 4 million
1650	Approx. 5.5 million
1750	Approx. 6.5 million
1800	9 million
1830	14 million
1850	18 million
1870	22.5 million

Data from *Encyclopaedia Britannica,* ninth and 1965 editions. Article *England.*

and entirely new types such as factories and railway stations began to appear. The rate of residential construction, and from the midnineteenth century on of schools and hospitals, was enormous. The total rate of construction has proceeded at an ever faster pace since 1750.

Table 8.2

Year	Population of Manchester and Salford	Average Annual Rate of Growth (%)
1588	Approx. 10,000	
		0.6
1757	19,839	
		2.3
1773	27,246	
		4.3
1783	More than 39,000	
		7.2
1801	89,752	
		6.2
1851	367,232	
		1.4
1881	517,649	

The emphasis also shifted from the traditional to the new. So far we have considered religious buildings, such as the Parthenon, the Pantheon, the Cathedrals of St. Denis and Beauvais, S. Pietro in Rome, St. Paul's in London, and the church of St. Geneviève in Paris. Only during the Roman Empire have other building types, such as baths and law courts, entered into our story. Churches continued to be built after the Industrial Revolution; indeed it is possible that more Christian churches were raised during the nineteenth century than in the preceding 1800 years, but none was so important from a technical point of view as the nonreligious structures of the same periods.

Domestic architecture for the common man, which had been largely a matter of self-help, became a professional activity. Many of the working-class houses in the new industrial cities, such as Manchester, were mass-produced by speculative builders. The amenities were deplorably poor by present-day standards but not necessarily by those of the time. Friedrich Engels, collaborator of Karl Marx, came to Manchester to work in his father's textile factory, first as an employee, then as a partner. He described the houses in 1844 in *Conditions of the Working Class in England.* Hogarth's paintings of housing in London before the Industrial Revolution show conditions that were different, inasmuch as the housing was not mass-produced as in Manchester, but no better.

Since the midnineteenth century many novels, social critiques, and architectural tracts have contrasted the industrial cities of early nineteenth-century England unfavorably with medieval villages. There is insufficient information on the rural life of the Middle Ages to prove or disprove this idealized picture, but evidence suggests that country life before the Industrial Revolution was less attractive than in the new cities. Surviving English villages in which the houses predate the eighteenth century, now so much admired for their charm, have a modern water supply, a modern sewage system, and paved roads that did not exist at that early time.

The Industrial Revolution is still in progress in the developing countries, and the same tendency to move to urban areas as in late eighteenth-century England is apparent. Housing conditions in the expanding cities are unsatisfactory, but life in the country is even less satisfying to those people who turn to urban living.

Comments by Victorian social reformers reinforce this view. George Godwin was born in 1813, the son of an architect. He, too, became an architect and in 1843, editor of *The Builder,* the leading architectural journal of its day. In *Town Swamps and Social Bridges* he described unsatisfactory living conditions in town and country (Fig. 8.8). Villages lacked paved roads, their houses were of poor quality, the water pump was frequently close to a cesspool, and sometimes the main water tank was built next to the village graveyard. Godwin pinpointed poor sanitation as the main concern, but it was not remedied until the late nineteenth century. The Victorians, however, tackled the problem with great energy and in the process transformed the environmental aspects of architecture (See Ref. 1.3, Section 7.5).

In the beginning life in the new industrial towns, as in the early industrial periods in Asia and Africa, was probably better than rural living. Growth of the cities created sanitary problems that did not exist in smaller communities. Rivers, such as the Irwell in Manchester, became open sewers. Gardens that had provided green space and a means of livelihood in times of unemployment disappeared. This deterioration and the periodic economic depressions caused riots quite early in the Industrial Revolution. In due course reforms gave working people a fairer share of the product of their labor, a political voice through parliamentary reform, the recognition of trade unions, and better housing brought about by the efforts of such bodies as the *Society for Improving the Conditions of the Labouring Classes,* founded in 1844, and the *Metropolitan Association for Improving the Dwellings of the Industrious Classes,* founded in 1841. The revolution that Engels predicted did not happen.

What then caused this tremendous change in the social conditions of England? Opinions differ, but most economic historians attribute it to a number of technical innovations between 1760 and 1840.

A Fever Village near Shrewsbury, from a Distance.

" Who would think it ?"

Part of the same Village, close.

" Who would doubt it ?"

8.8

George Godwin's pictorial comment on rural housing standards in the middle of the nineteenth century (Ref. 8.1, p. 62).

One was the development of the first spinning machine, patented by Sir Richard Arkwright in 1769. Arkwright entered into partnership with Jedediah Strutt, whose son William became one of the first designers of iron-framed buildings (Section 8.6). Arkwright's spinning machines were operated by water power, which made it necessary to assemble a large number of workmen in one factory. Spinning had always been done by families working in their own village homes, as it still is in many parts of Asia and Africa.

Water power was limited and some of it was already in use for flour milling and metal working. Steam power had been developed gradually during the seventeenth and eighteenth centuries, mainly for pumping the water out of deep mines; few mines were located near water on which a waterwheel could be operated. Thomas Savery obtained a patent for a steam pump in 1698 "for raising water by fire," and the first of the more successful engines of Thomas Newcomen was installed in 1712. It included an internal jet for condensing the steam and an automatic valve gear; Newcomen's engine worked at atmospheric pressure.

James Watt's invention therefore was not the steam engine as such but an external condenser, a steam-tight stuffing box for the piston rod, and a steam jacket to keep the cylinder hot. This enabled the engine to work over a greater temperature range and increased its efficiency considerably. It also meant that iron was needed for the cylinder. Watt entered into partnership with Mathew Boulton and established his works in Birmingham. It is a matter of opinion whether the steam engine was an integral part of the Industrial Revolution, for in 1840 more power was still being generated by waterwheels than by steam engines. Thermal power, however, was essential to industrialization thereafter and iron was needed for at least some of the engine parts.

The development of the iron industry had been held back by a shortage of timber for making charcoal. In 1709 Abraham Darby produced the first coke which made new supplies of fuel available for producing iron. In 1783 Henry Cort developed a rolling mill for rolling iron sections and in 1784 invented the puddling process by which molten pig iron was stirred with a long bar in a reverberatory furnace. This greatly reduced the cost of wrought iron. In 1775 John Wilkinson devised a boring mill which made it possible to bore engine cylinders to the fine limits of accuracy required in Watt's steam engines.

Transport also improved during the eighteenth century. Canal construction in England had lagged behind the Continent, but progress was rapid, and by the early nineteenth century a good network of canals for the transport of heavy goods was completed. For the first time since Roman days good roads were being built in England, but heavy transport continued by water. The first railway line (except for tracks in mines) was built between Stockton and Darlington in northern England in 1825; it was an immediate success and railways were soon in operation all over Europe and America. They had a profound effect on structural design (see Ref. 1.3, Section 2.1). The first steamship appeared in 1807 on New York's Hudson River, but it took a long time to displace the sailing ship.

Opinions also differ on the reasons that brought about the Industrial Revolution first in England rather than in prerevolutionary France, which in the mideighteenth century was a country with a greater population and greater wealth. In France a large part of the surplus capital was absorbed by the army and an expensive court, which was not, as in England, subject to parliamentary financial control. Labor was more mobile in England, where restrictive practices, such as serfdom and the guild system, had virtually disappeared. In an age of import restrictions England had the larger market because its greater colonial empire more than made up for the smaller home population. Better inland transportation and a better merchant navy may also have been contributing factors.

We shall now examine one aspect of the Industrial Revolution, the growth of the iron industry and the drop in the price of iron that gave England a great advantage over France. Possession of the right material proved more important than possession of a good theory.

8.6   EARLY IRON STRUCTURES

There are three main forms of iron: *Wrought iron,* which is almost pure iron, melts only at a temperature higher than any attainable before the nineteenth century. It is ductile at ordinary temperatures and easily forged when hot.

Steel, which contains 0.1 to 1.7% of carbon, has a lower melting point and is relatively soft when cooled slowly but extremely hard when quenched (i.e., cooled by plunging the hot metal in cold water). It can be forged at red heat.

Cast iron, which has a carbon content of 1.8 to 4.5%, is easily melted and cast into complicated shapes. It is a brittle material with little ductility.

Although cast iron had been produced in China since the sixth century B.C., in Europe and the Middle East until the Middle Ages all iron was produced as wrought iron directly from the ore. The ore was heated in a small furnace over a charcoal fire in which bellows raised the temperature, but the product was a soft sponge that never melted. It was then compacted and forged into the desired shape at white heat.

In Europe and the Middle East steel was made from wrought iron by prolonged heating in a charcoal fire which introduced carbon into the iron. The nature of steel as an alloy of iron and carbon was not understood, however. Aristotle knew that bronze was an alloy of copper and tin, but he considered steel to be a particularly pure form of iron, purified by fire. The making of good steel was a great art that depended on skill and experience, and it flourished in ancient India (see Section 4.2). A different method was perfected by the Persians and the Arabs. The textured metal of the famous Damascus sword blades was produced by a laminated structure of iron and iron carbide, obtained by forging at a relatively low temperature. Steel of comparable quality was not produced in Europe until the end of the eighteenth century. The tall blast furnace in which iron ore was mixed with charcoal in alternate layers was invented in the fifteenth century. The result was an iron with a much higher carbon content because of the intimate contact between the carbon and iron ore. Unlike wrought iron, this cast iron melted at temperatures that were attainable at the time with bellows. It could then be run off as liquid cast iron; the rocky parts of the ore formed a slag that floated on top.

Two problems impeded further progress. One was the limited supply of charcoal due to the shortage of timber (see Section 7.8); the other was the low strength of charcoal that limited the height to which alternate layers of iron ore and charcoal could be built without crushing the charcoal.

8.9

A view of the Upper Iron Works at Coalbrook Dale in 1758. The early factories fitted into the rural landscape. Only the chimneys indicate the presence of industry.

Coal had been mined on a small scale since early times. The Chinese used it in 1000 B.C. and the ancient Greeks and Romans used it on a small scale. Before the invention of satisfactory chimneys the fumes produced by the burning coal made it objectionable for heating and cooking. In the late fifteenth century it was used for brick making and in the sixteenth for glass and for smelting nonferrous metals. Attempts to use it for the manufacture of iron were unsuccessful until 1709 when Abraham Darby succeeded in reducing coal to coke by burning off the coal gas in an oven (Fig. 8.9).

Coke, which is mainly pure carbon, could support heavy loads of iron ore, and blast furnaces increased in size and efficiency. Cast iron, which had been a minor part of the total iron production, now took a more prominent role and was the principal material used in the first iron structures.

The new process greatly reduced the cost of iron and also increased its supply. Some wrought iron was produced by refining cast iron; that is, burning off most of the carbon.

A cheaper process was invented in 1784 by Henry Cort, who owned an iron works near Plymouth in England. The puddling furnace was a reverberatory furnace; that is, a furnace in which the ore was exposed to the action of the flame but not in contact with it. The heated gases were reverberated (i.e., deflected) by the sloping furnace roof onto the metal. Thus the iron could not absorb carbon from the fuel. The puddling process greatly improved the supply and reduced the cost of wrought iron, which was needed for machines and tools before steel became available in quantity.

Wrought iron could be welded simply by heating the parts to a white heat, placing them in contact, and rapidly hammering them together.

In 1783 Henry Cort obtained a patent for grooved rollers with which he could squeeze the metal into shape. He was soon producing plates, bars, and rods and other shapes followed. Cort had intended to install one of the new steam engines made by Boulton and Watt but found that he would be unable to obtain sufficient rotary power; therefore his first rolling mill used water power.

Steel remained a scarce material, but in the sixteenth century its heat treatment became more of a craft and less of an art. It had been known for centuries that steel could be hardened by rapid cooling in water (called quenching) and subsequently toughened by reheating and slow cooling (called tempering). Tempering produces distinctive colors, of which the dark blue corresponding to 295°C is the best known. In the seventeenth century these colors were classified and this made heat treatment easier to control.

Robert Hooke speculated in 1665 on their significance in his *Micrographia:*

Only I must not omit, that we have instances also of this kind even in metaline Bodies and animal; for these several colours which are observed to follow each other upon the polish surface of hardened steel, when it is by a sufficient degree of heat tempered or softened, are produced from nothing else but a certain thin Lamina of a vitrum or vitrified part of the Metal, which by that degree of heat, and the concurring action of the ambient Air, is driven out and fixed on the surface of the steel.

And this hints to a very probably (at least, if not true) cause of the hardening and tempering of steel, which has not, I think, been yet given, nor, that I know of so much as thought of by any. And that is this, that the hardness arises from a greater proportion of a vitrified substance interspersed through the pores of the steel. And that the tempering or softening of it arises from the proportionate or smaller parcels of it left within those pores. These will seem the more probable, if we consider these Particulars.

First. That the pure parts of Metals are of themselves very flexible and tuff; that is, will endure bending and hammering, and yet retain their continuity.

Next, that the Parts of all vitrified Substances, as all kinds of Glass, the Scoria of Metals, &c. are very hard, and also very brittle, being neither flexible nor malleable, but may by hammering or beating be broken into small parts or powders. . . .

Fifthly, That Iron is converted into Steel by means of certain salts, with which it is kept a certain time in the fire.

This is a remarkably percipient observation, except that Hooke failed to identify the carbon derived from the charcoal as the critical constituent of steel; instead he looked for a "salt."

Steel, however, remained a costly material until Bessemer found a new process in 1856. In the meantime the choice was between the ductile, but not very strong, wrought iron and the brittle cast iron which had a high compressive strength.

We have noted that iron had been used occasionally by the Greeks (Section 3.2), and that the Romans used it for cramping masonry (Section 3.9). It was sometimes used for ties in Gothic construction to absorb the horizontal thrust of a vault (Section 6.5) and quite commonly in large Renaissance structures for the same purpose (Sections 7.2, 7.3, and 7.6).

The first structure in which iron was the main structural material was the bridge over the Severn at Coalbrook Dale, twelve miles downstream from Shrewsbury, constructed in a mere three months in 1779. Thomas Tredgold, who in 1824 published a book on the structural use of cast iron, described its genesis:

One of the boldest attempts with a new material, was the application of cast iron to bridges; the idea appears to have originated, in the year 1773, with the late Thomas Farnolls Pritchard, then of Eyton Turret, Shropshire, architect, who in communication with the late Mr. John Wilkinson, of Brosely and Castlehead, ironmaster, suggested the practicability of constructing wide iron arches, capable of admitting the passage of the water in a river, such as the Severn, which is much subject to floods. This suggestion Mr. Wilkinson considered with great attention, and at length carried into execution between Madely and Brosely, by erecting the celebrated iron bridge at Cole Brook Dale, which was the first construction of that kind in England, and probably in the world. This bridge was executed by a Mr. Daniel Onions, with some variations from Mr. Pritchard's plan, under the auspices, and the expense of Mr. Darby and Mr. Reynolds, of the iron works of Coal Brook Dale. Mr. Pritchard died in October, 1777. He made several ingenious designs, to show how stone or brick arches might be constructed with cast-iron centres, so that the centre should always be a permanent part of the arch. These designs are now in the possession of Mr. John White, of Devonshire Place, one of his grandsons, to whom I am indebted for the preceding particulars (Ref. 8.17, pp. 9 and 10).

The original design for a masonry bridge with a fixed cast-iron centering and the final design of an all-cast-iron bridge are shown in Fig. 8.10. It is not known what design calculations, if any, were performed. The bridge has been very successful, however.

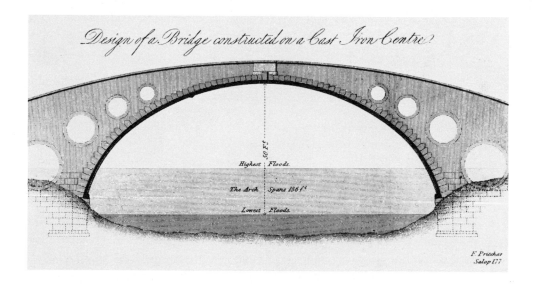

8.10a

The first Iron Bridge at Coalbrook Dale, 1779. T. F. Pritchard's design for a masonry bridge with a fixed cast-iron centering.

It carried heavy traffic for more than a century and a half and is still in use as a footbridge. There has been no significant damage to its structure in two centuries. Insufficient provision was made for the horizontal pressure exerted by the arch and in 1969 remedial works on the abutments became necessary (Ref. 8.3).

The Coalbrook Dale bridge has a span of 30 m (100 ft). The joints are reminiscent of those in timber structure, but the arch as a whole looks almost like a masonry arch in which the mortar joints are replaced by cast iron and the masonry is replaced by air. The bridge hardly interferes with the view of the landscape behind it and thus gives an appearance of lightness compared with the traditional masonry arch. There are five parallel semicircular arches with a total weight of 378 tons. The ribs were cast in the Coalbrook Dale Iron Works in two parts, each about 21 m (70 ft) long. Steinman (Ref. 8.4, p. 132) disagrees with Tredgold's account and credits Abraham Darby III, the ironmaster, with the design. In 1787 Darby received a gold medal for it from the Royal Society of Arts.

The second iron bridge, designed by Thomas Telford, subsequently the first president of the Institution of Civil Engineers to whom Tredgold dedicated his book on cast iron (Ref. 8.17), was erected only three miles from Coalbrook Dale.

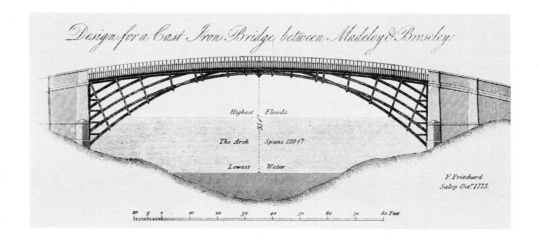

8.10b

T. F. Pritchard's design for a cast-iron bridge at Coalbrook Dale.

The third iron bridge was completed in 1796 at Wearmouth, near Sunderland, in the north of England. It has a span of 71 m (236 ft) and a ratio of span to rise of 7 (compared with 2 for Coalbrook Dale). The castings were originally intended for a 400-ft span over the Schuylkill River in Philadelphia, designed by Thomas Paine, better known as a revolutionary and author of the *Rights of Man*. They were made to Paine's specification at a foundry in Rotherham, Yorkshire, and exhibited with the models in London. Then Paine's financial arrangements collapsed and he became

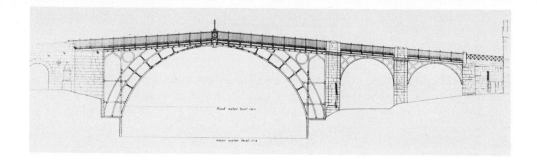

8.10c

This all-cast-iron bridge at Coalbrook Dale was actually built and is still standing.

engrossed in the French Revolution. The castings were reclaimed by the manufacturer. Lewis Gordon, the first professor of engineering at Glasgow University, named Rowland Burdon as the designer of this bridge (Ref. 4.6, Vol. IV, p. 334). Burdon, a local member of Parliament, raised the money for its construction. He had also practiced as an architect and engineer.

In the late eighteenth century it was decided to replace London Bridge (Section 5.1) whose nineteen arches were causing a severe obstruction to the flow of the river. A proposal was submitted by Thomas Telford in 1801 to span the river with a single cast-iron 183-m (600-ft) arch. The Committee felt unable to pronounce on this scheme and appointed a panel of professors of mathematics from Oxford, Cambridge, and Edinburgh, two other mathematicians, three ironmasters, and John Rennie, a civil engineer. Telford's light, low-rise arch was ahead of its time and the panel rejected it. Eventually John Rennie received the commission and between 1825 and 1831 built a masonry bridge consisting of five semielliptical arches. This structure was recently demolished and reerected in Arizona.

The reason for using iron in bridges was the greater span allowed by the relatively low weight of the structure. In architecture it commended itself particularly because, unlike timber, it did not burn. Its first use was in France, but the main development of the iron-framed building took place in the new factories of England.

Hollow pots, already used by the ancient Romans (Fig. 3.24), were reinvented in France in 1785 by the engineer-architect Eutashe de Saint-Fair and were used in England shortly after. Mathew Boulton (of Boulton and Watt) wrote in 1793 to William Strutt:

I understand you have some thoughts of adopting the invention of forming Arches by means of hollow pots and thereby saving the use of Timber in making Floors, and guarding against Fire. Allow me to say that I have seen at Paris floors so constructed, and likewise at Mr. George Saunders' in Oxford St., London, who is an eminent Architect (Ref. 8.5, p. 184).

Open-web wrought iron girders were also invented in France. Their first use was in the Théâtre-Français, built in Paris between 1787 and 1790 and probably designed by Victor Louis who built a number of theaters. The roof was of wrought iron, with two ties to absorb the horizontal thrust, so that the loadbearing walls were quite thin. A number of diagonals stiffened the structure, but it is not known whether any calculations were made (Ref. 8.18, p. 174).

The declared objective in this case was to produce a building more fireproof than one with a timber roof and floors. This was also the main reason for using iron in England. The buildings most at risk were the new textile mills. A typical mill had to be five or six stories high in order to bring all the machinery as close as possible to the source of power, which was a waterwheel (Fig. 8.14) or a steam engine. The power was transmitted from this prime mover to shafts and from the shafts to the machines by further belting. The longer the belting, the greater the loss of power. This method of power transmission had low efficiency and thus generated an appreciable amount of heat which was a fire hazard. The cotton presented another hazard, and there were many fires in the timber-framed mills. It was not uncommon, however, to span across the entire width of the mill, up to 28 ft (8.5 m), with timber beams about 12 in. (305 mm) square. Some of these mills still survive in Derbyshire, where Arkwright and Strutt built their first cotton mill (Section 8.5).

"Fireproof" structures were first built in 1792 by William Strutt, son of Arkwright's partner, who substituted brick jack arches for the timber floor and protected the timber beams with plaster below and tile above. To absorb the horizontal thrust of the brick arches he used wrought iron ties (Fig. 8.11). This greatly increased the weight of the floor, and he inserted cast-iron columns, generally at 9 ft (2.7 m) centers, to reduce the span. Arches of hollow pots were used as a ceiling on the top floor to protect the timber roof.

Cast iron does not soften so readily in fire as wrought iron, and this form of construction stood up well in factory fires.

In 1803 Strutt built his first mill with a complete iron frame. The cast-iron girders had a flange at the base to serve as a support for the brick jack arches (Fig. 8.12). This flange also improved their strength, for it provided additional material on the tension face; cast iron is much stronger in compression than in tension. It is not known, however, whether this structural advantage was appreciated at the time.

Numerous factories were built in England during the nineteenth century, and cast-iron framing became the normal form of construction until it was replaced by steel. In due course the technique spread to the Continent and the United States. The columns used in Strutt's earliest mills were solid, but in time they were made hollow to give them greater stiffness for the same amount of material. Cast iron, because of its low tensile strength, was not entirely suitable for use in beams and wrought iron beams were sometimes used in conjunction with cast iron columns. It was an excellent material for columns, and these continued to be made until 1914 when World War I stopped their manufacture; it was not resumed after the war.

The age of cast iron overlapped with the classical revivals in architecture (see Section 8.8), and in due course cast-iron columns with Doric, Ionic, and particularly Corinthian decorations appeared (Fig. 8.13). These columns were much slenderer than the Greeks had ordained, but they had a charm of their own. They survive in old railway stations, in some commercial shopfronts in England and America, and in private houses, both large and small, in the subtropical region of America, notably in New Orleans, and the older towns of Australia (Fig. 8.13).

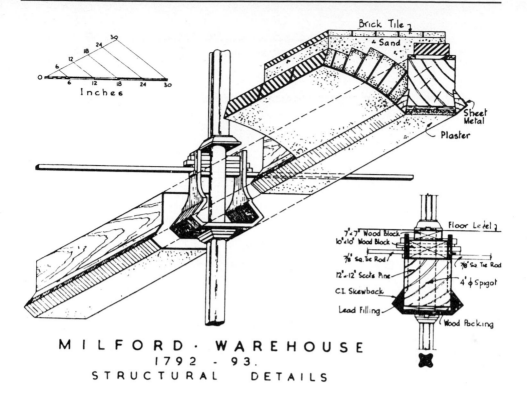

Brick Tile

Sand

Sheet Metal

Plaster

Inches

Floor Level

7"×7" Wood Block

10"×10" Wood Block

7/8" Sq. Tie Rod

7/8" Sq. Tie Rod

12"×12" Scots Pine

4' φ Spigot

C.I. Skewback

Lead Filling

Wood Packing

M I L F O R D · W A R E H O U S E
1 7 9 2 - 9 3.
S T R U C T U R A L   D E T A I L S

8.11

"Fireproof" structure consisting of brick jack arches, tied with wrought-iron bars, timber beams, protected by plaster underneath and brick tiles above, and cast-iron columns (Ref. 8.5, p. 187).

Cast iron was used in 1818 by John Nash in the Royal Pavilion at Brighton, the lighthearted seaside home of the Prince Regent. The kitchen had very tall and slender cast-iron columns that bore huge capitals of palm leaves. The drawing room had columns equally slender, but not quite so tall; their capitals were lotus leaves. This was one of the few examples of the use of cast iron as a structural material in a prestige building.

The industrial cast-iron buildings were designed by engineers not architects. Unlike modern factories, they were planned around the power plant (Fig. 8.14):

The whole motion of the machinery is taken from the great waterwheel, situated underneath the wing, in the lowest room in the mill; and as it is of so great a size, namely 18 feet diameter and 23 feet long (5.4 × 6.9 m), that no cast-iron girder could be thrown across it strong enough to support the arches for the wing above it, a strong stone arch is built over the wheel; and to resist the thrust of this arch, two strong iron bolts are extended across it, and render it as strong

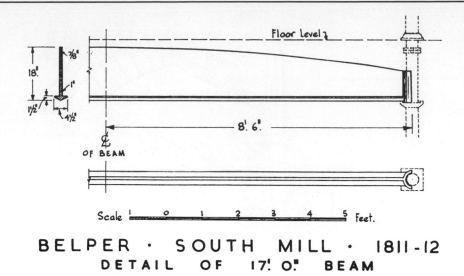

BELPER · SOUTH MILL · 1811-12
DETAIL OF 17′ 0″ BEAM

8.12

Cast-iron frame for a "fireproof" structure. The floor was formed by brick jack arches, as in Fig. 8.11 (Ref. 8.5, p. 199).

as possible, so that the iron columns may be raised upon it as safely as they could upon foundation piers, like the others, but as a precaution against overloading the walls which, as they include the water-wheel, would ruin everything if they settled, the arches of the wing immediately over the wheel are built, instead of solid brick, with small pots like garden pots, so that they are light, but sufficiently strong to bear anything which is ever required to be loaded upon them. (From a description in *Rees' Cyclopaedia*, London, 1819, quoted by Johnson and Skempton, Ref. 8.5, p. 195).

The power plant and the factory building were commonly designed and built by the same engineering firm. The firms that engaged in this sort of work, of whom Boulton and Watt is the best known, soon acquired experience in structural design and developed suitable methods of calculations. Skempton (Ref. 8.6, p. 179) discussed the theory and pointed out that the calculations were frequently supplemented by full-scale tests. It would not have been unduly difficult to build a test frame in an engineering works and use pieces of iron for loading.

The first textbook on the design of cast iron, published by Thomas Tredgold in 1824, remained a standard work for several decades. Its most important feature was a table extending over fourteen pages on the strength and stiffness of square beams in

8.13

Cast-iron columns and balcony railings in Edgecliff Road, Woollahra, an inner suburb of Sydney, Australia.

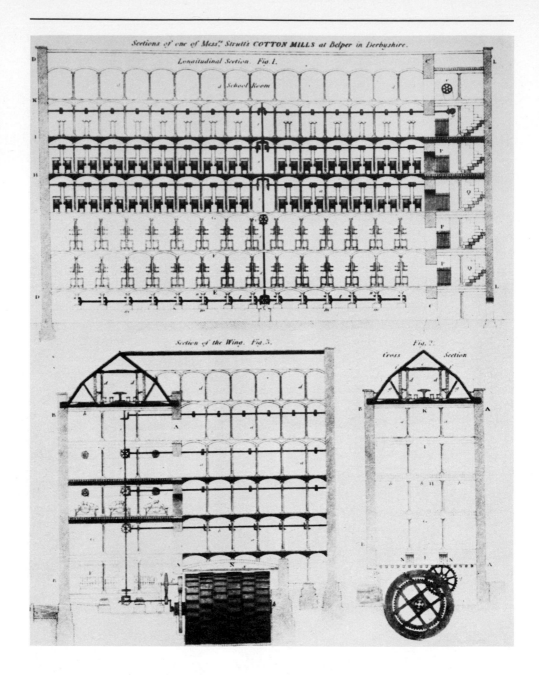

8.14

Section through Belper North Mill in Derbyshire, designed by William Strutt in 1803. From a drawing by John Farey in Rees' *Cyclopaedia*, London, 1819; article *Manufacture of Cotton*.

The Age of Reason and the Industrial Revolution

terms of various depths and spans and another for the safe load of circular columns in terms of height and column diameter. These tables are not so useful as modern safe-load tables for steel, but they mark a first step in that direction.

Tredgold, like many British engineers in the nineteenth century, started his career as a carpenter and was not enamored of science:

The manner in which the resistance of materials has been treated by most of your common mechanical writers, has also, in some degree, misled such practical men as were desirous of proceeding upon surer ground; and has given occasion for the sarcastic remark "that the stability of a building is inversely proportional to the science of the builder" (Ref. 8.17, pp. 2–3).

He had a particular dislike for the use of calculus (called fluxions by Newton and subsequent English writers), which had by that time become common procedure in French papers on structural theory:

I have avoided fluxions in consequence of the very obscure manner in which its principles have been explained by the writers I have consulted on the subject. I cannot reconcile the idea of one of the terms of a proportion vanishing for the purpose of obtaining a correct result; it is not, it cannot be good reasoning (Ref. 8.17, p. ix).

Instead he used the sum of an infinite series to obtain the term in the theory of bending which we now call I (second moment of area). He did, however, obtain the correct answer for Galileo's problem; that is, the cantilever carrying a concentrated load at its end (Ref. 8.17, p. 125).

$$W = \frac{fbd^2}{6L} \tag{8.16}$$

The above equation is given in Tredgold's own notation, and it is interesting to note that the same notation is still used today. Tredgold's "neutral axis" appears to be the first time that the term was used in English. Since his book appeared two years before Navier's Leçons (Section 8.3), he was evidently familiar with the theory of his time.

In spite of his remarks about science, Tredgold, frequently quoted French writers, but his main source of theory was Lectures on Natural Philosophy by Thomas Young, who in turn derived his bending theory from Coulomb (Section 8.3).

Tredgold's most important contribution was the solution to the problem of columns. Euler had derived the equation for elastic buckling in 1757 (Section 8.4), but it applied only to very slender columns (Fig. 8.7). Practical cast-iron (steel, timber, or reinforced concrete) columns are much stubbier, but there may still be some buckling, particularly in the elegantly thin cast-iron columns popular in the early nineteenth century. Tredgold's formula, which is essentially empirical, is of the type

$$P = \frac{fA}{1 + aL^2/d^2} \tag{8.17}$$

where P is the safe load for a circular column, f is the safe column stress, A is the cross-sectional area of the column, L is its length, d is the diameter, and a is a

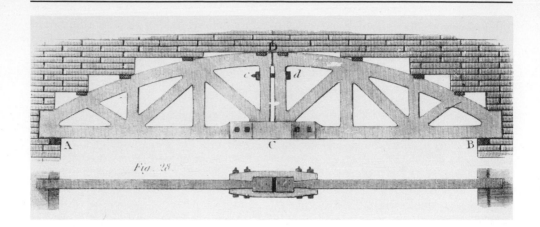

8.15

Tredgold's "sketch for a beam to bear a considerable load distributed uniformly over its length, when the span is so much as to render it necessary to cast in two pieces. The connection may be made by a plate of wrought iron on each side of C, with indents to fit the corresponding parts of the beam. Wrought iron should be preferred for the connecting plates, because being more ductile it is more safe" (Ref. 8.17, p. 296).

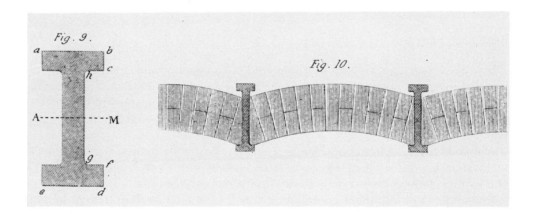

8.16

Tredgold's cast-iron floor beams: "Fig. 9. The strongest form of section for a beam to resist a cross strain. AM is called the neutral axis. Fig. 10. Shows an application of the section of Fig. 9 to form a fireproof floor, the projection serving the double purpose of giving additional strength, and forming a support for the arches" (Ref. 8.17, p. 290).

constant. This is now known as Gordon's formula, after the first Professor of Engineering at Glasgow University, and still widely used in the early twentieth century; Tredgold, however, published it before Gordon.

His book included a detailed review of experiments carried out in France and England, many of the latter in engineering works.

Among the plates is an illustration of a truss that is reminiscent of carpentry (Fig. 8.15) and another of a floor consisting of cast iron *I*-beams and brick jack arches (Fig. 8.16). Tredgold noted that the resistance of the section was directly proportional to the square of the distance of its parts from the neutral axis. He then observed that the parts should be of equal thickness because the section "is sometimes even fractured by irregular cooling. . . . The form of section which I usually adopt in order to fulfill these conditions, is represented in Fig. 9" (Fig. 8.16 in this book) (Ref. 8.17, p. 55).

In the early nineteenth century British engineers therefore had sound practical design methods, even though their theoretical basis was primitive, and ample opportunity to use them in the numerous cast-iron framed factories built during that time. Their supply of cast iron was generous. These advantages were more important than an excellent French engineering education and the better theory developed by French engineers. When peace returned to Europe after the Napoleonic wars, the importance of a mathematically based theory became evident (Ref. 1.3, Section 2.5).

8.7   THE RENAISSANCE OF CONCRETE

Concrete had been the principal building material of Imperial Rome; it regained that role only in the twentieth century. During the Middle Ages concrete continued to be used as a foundation material and as filling in the cores of massive piers, but it was generally weak compared with Roman and modern times. Interest in concrete, as in all things Roman, was renewed during the Renaissance. Brunelleschi studied it. Bramante stated that he used concrete in the ancient manner for the cores of his piers in S. Pietro, later enclosed in Michelangelo's larger piers.

A full-scale revival was held back by the failure to understand the chemical basis of Roman concrete (Sections 3.5 and 3.6). The Romans had been fortunate; they had an ample supply of labor which enabled them to use a very dry mix and compact it thoroughly. In addition, they had in most parts of Italy, and specifically near Rome, a supply of volcanic sand that interacted with the lime to form a material with some resemblance to modern cement. When this was not available, they used powdered brick, of which they had plenty from demolitions.

The reasons for the ingredients of Roman concrete were, however, not understood. Vitruvius, whose *Ten Books* had been studied carefully since the fifteenth century, trailed a red herring by his glowing description of the effect of pozzolana on concrete (Section 3.5). Supplies of pozzolana, a natural volcanic ash, were still available in Italy, and it was recognized that another volcanic ash, known as Rhenish trass (or tarras), found near the upper Rhine River and marketed in the Dutch ports, had similar properties. These materials were made expensive by the cost of transport and a modern concrete industry could not be built on the limited supply. In particular, there were no suitable deposits in England. Vitruvius also stated:

With regard to lime we must be careful that it is burned from a stone which, whether soft or hard, is in any case white. Lime made of close-grained stone of the harder sort will be good in structural parts, lime of porous stone in stucco. After slaking it, mix your mortar, if using pitsand, in the proportions of three parts of sand to one of lime; if using river or sea-sand, mix two parts of sand with one of lime. These will be the right proportions for the composition of the mixture. Further, in using river or sea-sand, the addition of a third part composed of burnt brick, pounded up and sifted, will make your mortar of a better composition to use (Ref. 2.3, Book II, Chapter 5, p. 45).

The part of this passage often repeated was the insistence on white lime. Even Bernard Forest de Belidor in *Architecture hydraulique,* a book with a strong scientific bias published in 1753, wanted lime to be pure white and made from strong limestone. It is natural, but not correct, to assume that the best lime for mortar is the purest and whitest. Vitruvius' recipe worked because he specified the use of *fossiciae* (a sand of volcanic origin containing some alumina, translated by Morgan as pitsand) or the addition of powdered brick.

The use of volcanic sand or powdered brick as an admixture to mortar to improve its strength and resistance to water did not completely die out. Haegermann (Ref. 3.34, p. 27) quoted a few examples of their use, but these were isolated instances, and the water-resistant qualities of these structures are open to question.

When the first modern lighthouses were built on rocks that were sometimes awash, it was generally considered that lime mortar would not withstand the action of the water and that masonry consequently could not be used. One of the earliest of these lighthouses was built in 1699 on the Eddystone, a rock off the coast of Plymouth, England, which lay on the main shipping route from London to New York and Boston. This timber structure was destroyed in 1702 by a spring tide. The second lighthouse, also of timber, built in 1706, was destroyed by fire in 1755. The Brethren of Trinity House, a fraternity of seamen founded by Sir Thomas Spert, Henry VIII's Controller of the King's Navy, had in the sixteenth century been granted powers over all navigation lights. They then appealed to the Earl of Macclesfield, President of the Royal Society of London, for advice. He recommended John Smeaton, then 31 years old, who had been elected a Fellow two years before.

Smeaton built the third Eddystone lighthouse between 1756 and 1759 (Fig. 8.17a). He decided that only a masonry structure with a water-resistant mortar would be able to survive the wave action but had the blocks of Portland stone cut with interlocking dovetail joints (Fig. 8.17b). He then set out to find a reliable "water" or "hydraulic" cement, that is, a cement that, unlike lime, was not water-soluble and could thus be used for hydraulic works.

He experimented with lime made from various deposits and first of all disproved the dictum of Vitruvius and Belidor that the strongest limestone made the strongest lime; he found no difference in strength between lime made from soft white chalk and the hardest white marble.

In the course of this investigation he observed that the strongest limes were produced from limestones that were not white; he obtained a particularly strong lime mortar from a bluish limestone from Aberthaw. On the advice of a chemist named Cookworthy he then analyzed this rock by dissolving the lime slowly with *aqua fortis* (concentrated nitric acid):

South ELEVATION of the STONE LIGHTHOUSE completed upon the EDYSTONE in 1759.
Shewing the Prospect of the nearest Land, as it appears from the Rocks in a clear calm Day.
Engraved in the Year 1763, by Mr. Edwd Rooker, The Figures by Mr. Saml Wale.

8.17a

The third Eddystone lighthouse, built by John Smeaton from 1756 to 1759.

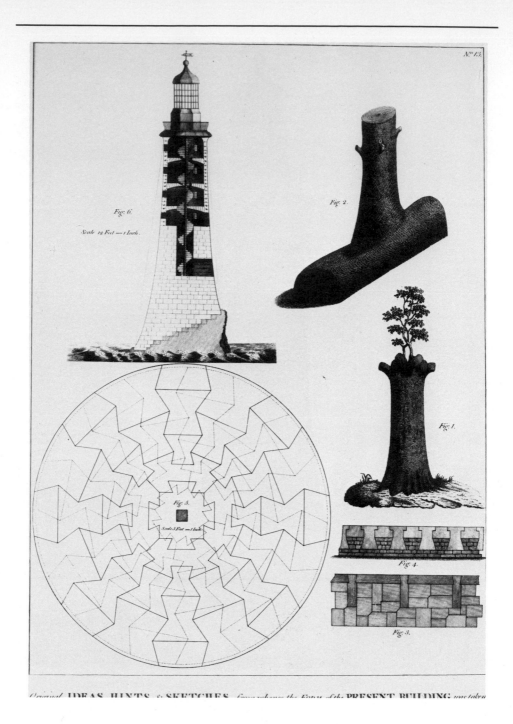

Original IDEAS HINTS & SKETCHES from whence the Form of the PRESENT BUILDING was taken

8.17b

The concept of the Eddystone lighthouse and the interlocking masonry blocks (from Smeaton's *Description*, Ref. 8.14).

On trying Aberthaw lime in this way, it was dissolved in the aqua fortis; but the solution appeared very dark and muddy, and on examination I found a small quantity of undissolved sandy particles at the bottom, some of them transparent like crystals, but mostly very minute, and of dirty appearance. The muddy residuum being brought into an argillaceous state, was very tough and tenacious while soft; and when sufficiently hardened, being worked into a little ball, and dried, in that state it appeared to be a very fine compact clay, weighing nearly one-eighth part of the original mass. One of the balls having been burnt became a good compact brick, which being of a reddish colour, it from thence appeared, as I was told was the case, that it had an admixture of iron in its composition. . . .

195. From the experiments now related, I was convinced that the most pure limestone was not the best for making mortar, especially for building in water. . . . The most pure lime affording the greatest quantity of Lime Salts, or impregnation, would best answer the purpose of Agriculture: whereas, for some reason or other, when a limestone is intimately mixed with a proportion of Clay, which is by burning converted into Brick, it is made to act more strongly as a Cement (Ref. 8.14, Book III, Chapter 4, pp. 107–108).

Smeaton used a mortar made with lime from Aberthaw limestone with an addition of Italian pozzolana. The third Eddystone lighthouse never gave any trouble, although in 1882, after 123 years, it had to be rebuilt on another rock because the rock under Smeaton's lighthouse had been eroded by wave action.

The Eddystone lighthouse, greatly admired as an engineering feat, established Smeaton's reputation. In his old age he wrote a description of its construction (Ref. 8.14), which appeared in 1791, a year before his death (Fig. 8.17a). In it he gave an account of his investigations on the strength of mortar. On the cover is a view of the lighthouse completely covered by a huge wave (Fig. 8.18), rather more dramatic than the reality.

In his book Smeaton stated his belief that there were many deposits of limestone in England that could be used to produce hydraulic lime. The first successful product was patented in 1796 by James Parker who obtained his hydraulic lime by burning "certain stones of clay or concretions of clay containing veins of calcareous matter" found at Northfleet, Kent, not far from London, the most important market. Smeaton had, in the traditional manner used at least since medieval times, first produced quicklime and then slaked lime; that is, he burned the lime only enough to drive off the carbon dioxide and then added water. Parker used a much higher temperature but not enough to vitrify the materials. He then ground the product into a powder which he called *Roman cement*. Actually the material was quite different from that used by the Romans; indeed, because of the ready availability of the raw materials, it represented a distinct advance. However, ancient Roman concrete still had a great reputation.

We now call Parker's product a natural cement because it was made by burning the raw material as found, without mixing two separate materials. Also in 1796, Lesage, a French military engineer, produced a natural cement from material found in Boulogne. Cement manufacture soon spread to other parts of England and the Continent. In the United States the first natural cement was produced in 1818. Natural cement remained an important material throughout the nineteenth century, but by 1900 it had been overtaken by portland cement.

The first portland cement was made in 1811 by Joseph Aspdin (Ref. 8.7) who started work as a bricklayer. The patent taken out in 1824 described the process:

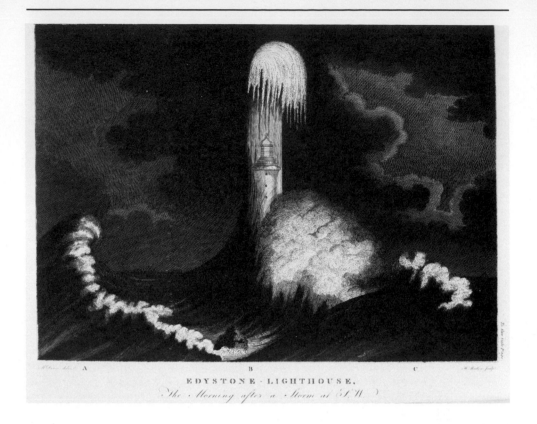

EDYSTONE · LIGHTHOUSE,
The Morning after a Storm at S.W.

8.18

The frontispiece of Smeaton's *Description* (Ref. 8.14).

I take a specific quantity of limestone, such as that generally used for making or repairing roads, and I take it from the roads after it has been reduced to a puddle, or powder; but if I cannot procure a sufficient quantity of the above from the roads I obtain the limestone itself, and I cause the puddle or powder, or the limestone, as the case may be, to be calcined. I then take a specific quantity of argillaceous earth or clay and mix them with water to a state approaching impalpability, either by labour or by machinery. After this proceeding I put the above mixture into a slip pan for evaporation, . . . until the water is entirely evaporated. Then I break the said mixture into suitable lumps, and calcine them in a furnace similar to a lime kiln until the carbonic acid is entirely expelled. The mixture so calcined is to be ground, beat, or rolled to a fine powder and is then in a fit state for making cement or artificial stone (Ref. 8.17, p. 19).

This is an artificial cement made by mixing together two separate materials. The materials are available in unlimited quantities in almost every country, and portland cement has now become the cheapest of all manufactured products per unit weight.

The name was probably inspired by Smeaton who wrote in his book on the Eddystone lighthouse that his mixture of Aberthaw lime and Italian pozzolana formed "a cement which would be equivalent to the best commercially obtainable Portland stone for strength and durability." Aspdin owned a copy of this book. Portland cement looks, of course, quite different from the white Portland stone, a fine-grained limestone used for the Eddystone lighthouse and also for some of the most important buildings erected in London during the eighteenth and nineteenth centuries (Section 8.8). Portland-cement concrete, however, is as strong and durable as Portland stone.

The material now known as portland cement is somewhat different from Aspdin's. Isaac Charles Johnson who has claimed to be its inventor started work at the age of 14, and at the age of 36, in 1837, became the works manager of J. B. White and Son, manufacturers of Frost's Cement, a material patented in 1822. Following the success of Aspdin's cement, Johnson's company instructed him in 1844 to produce a comparable product by any lawful means. Reminiscing at the age of 70, he wrote:

There were no sources of information to assist me, for although Aspdin had works, there was no possibility of finding out what he was doing, because the place was closely built in, with walls some 20 ft. high and with no way into the works, excepting through the office.

I am free to confess that if I could have got a clue in that direction I should have taken advantage of such an opportunity, but as I have since learned, and that from one of his later partners, the process was so mystified that anyone might get on the wrong scent—for even the workmen knew nothing, considering that the virtue consisted in something Aspdin did with his own hands.

Thus he had a kind of tray with several compartments, and in these he had powdered sulphate of copper, powdered limestone and some other matters. When a layer of washed and dried slurry and the coke had been put in the kiln, he would go and scatter some handfuls of these powders from time to time as the loading proceeded, so the whole thing was surrounded by mystery.

What did I then do? I obtained some of the cement in common use and, although I had paid some attention to chemistry, I would not trust myself to analyse it, but I took it to the most celebrated analyst of that day in London (*Building News*, December 1880).

The chemical analysis produced nothing useful and Johnson then undertook experiments with different proportions of chalk and clay. He discovered the advantage of a high firing temperature accidentally, for it had been customary to keep the temperature of burning below vitrification. He then experimented with some discarded clinker and found that it had superior strength. Thus he established the concept of burning the mixture at a white heat at which the materials begin to fuse. Johnson's claim to the invention of portland cement has been accepted by Professor Wilhelm Michaelis, a leading cement chemist of the nineteenth century.

Both Aspdin and Johnson, by then an independent cement manufacturer, exhibited their products at the Great Exhibition of London in 1851. The judge, General C. W. Pasley (who was Chief Inspecting Officer of Railways), praised both products but considered Aspdin's slightly better.

The invention of modern concrete was contingent on the discovery of a waterproof cement. A mixture of pebbles and lime had been used as a foundation material and as filler in thick walls and buttresses in many medieval buildings. If one credits Smeaton with the invention, it is purely in the sense that he employed a *hydraulic* mortar mixed with pebbles. He did so for the first time in 1760 in the construction of a lock for the River Calder in the English Midlands. Smeaton himself wrote that the idea of using hydraulic mortar mixed with pebbles for the lock came to him when he saw the ruins of the Saxon Corfe Castle in Dorsetshire whose walls were filled with pebbles set in lime mortar. According to the Oxford Dictionary, the word concrete, meaning ''a composition of stone chippings, sand, gravel, pebbles, etc., formed into mass with cement (or lime),'' was not employed before 1834.

The fast-growing production of cement was absorbed mainly by cement mortar used as a jointing material for bricks and masonry and for waterproof rendering (Section 8.8). Its use in concrete for harbor and river works remained on a small scale until the midnineteenth century.

The first really important use of concrete was in 1859 when the Metropolitan Board of Works employed it in the construction of the London Drainage Canal.

In architecture concrete advanced even more slowly in the eighteenth and nineteenth centuries. After the 1800s it was sometimes substituted for sand (Fig. 8.11) as a filler above the brick jack arches in fireproof construction.

The first precast concrete blocks, Ranger's Patent for Artificial Stone, were advertised in 1832 and are known to have been used in a number of buildings in London and Brighton in the 1830s.

Progress was a little faster on the Continent, which had a stronger tradition of *pisé* construction than England. Mud huts (Section 1.1) were part of the ancient tradition of building. They survive in Africa (Fig. 1.2), Asia, and South America and were probably in common use in Europe until a few centuries ago. In rainy England mud walling took the form of wattle-and-daub (Section 5.6), but in some parts of France *pisé de terre* was still used in the nineteenth century.

Jean Baptiste Rondelet described it in detail in his ten-volume *Traité de l'art de bâtir* (Treatise on Building Crafts), published in 1802, which became a standard work. The German poet Johann Wolfgang von Goethe wrote about it in 1819 in a rhyme in *Westöstlicher Divan:*

Getret'ner Quark
wird breit, nicht stark.
Schlagst Du ihn aber mit Gewalt
in feste Form,
nimmt er Gestalt.
Dergleichen Steine wirst Du kennen,
Europaer 'Pisé' sie nennen.

(If you walk about in mud with your feet, it spreads and has no strength; but if you beat it with force into formwork it assumes its shape. It becomes like stone; Europeans call it Pisé.)

Pisé was made by ramming unburnt clay or chalk in a damp condition into formwork without reinforcement; the formwork was then removed. Substitution of concrete for damp clay thus produced a concrete house, the first of which appeared in the 1830s; but in France as in England concrete established itself as an important material only after 1850.

Rondelet had written that the pisé walls became *une seule pièce.* François Coignet, who was one of the first to build concrete houses in the pisé manner, also drew attention to the advantages of concrete buildings made all in one piece and used the term *monolithe,* previously applied to stone columns or monuments cut from one piece of stone. The claim that concrete buildings were *monolithic* became one of the important arguments in its favor and also one of the most difficult structural design problems, which was solved only in the twentieth century (Ref. 1.3, Chapters 2 and 3).

8.8 ARCHITECTURE AND BUILDING CONSTRUCTION IN THE EIGHTEENTH CENTURY

The Renaissance created a completely new type of structure, and the Baroque introduced highly sophisticated geometry into construction. A glance at the illustration in Guarini's *Architettura civile* (Ref. 2.13) demonstrates the complexity of the design, even if it is not always apparent from the Baroque buildings, curving with effortless ease.

The architecture of the churches and palaces of the eighteenth century added little of scientific interest, but it developed a great skill in craftsmanship which is in many respects reminiscent of Imperial Rome, particularly in its ability and willingness to produce a good fake.

Pliny wrote toward the end of the first century A.D. that "the best woods for cutting into layers and for employing as veneers are the citron, the terebinth, the different varieties of maple, the box, the palm, the holly, the holm-oak, the root of the elder and the poplar." (Ref. 8.8, p. 125). Tortoiseshell was also cut into thin layers for veneering and glued to the timber. This was an entirely different matter from inlay, which was a much older custom; Homer had described Penelope's chair as inlaid with ivory and silver.

Veneered furniture ceased to be made with the fall of the Roman Empire. Medieval and Renaissance furniture was always solid, although sometimes exquisitely carved. Inlays again became common in the seventeenth century and veneering was widely used for quality furniture in the eighteenth.

There was a parallel development in architecture. Vitrivius (Section 3.4) described vaulting made with wooden furring strips and several coats of rendering. None of this construction survives.

In the eighteenth century rooms with vaulted ceilings were again constructed from lath and plaster. Some were exquisitely finished and are among the most tasteful interiors of all time.

The domes on eighteenth-century buildings were frequently formed by timber trusses with a nonstructural roof covering on the outside and a plaster ceiling on the inside. This is obvious in the little onion domes that John Nash placed on the Brighton Pavilion but surprising in the large baroque dome of Sir John Vanbrugh's Castle Howard in Yorkshire, built in the early 1700s.

The eighteenth century was also the first to fake its exteriors. Because London was built on clay surrounded by chalk, it had always been necessary to import stone. By 1600 Portland stone imported from the coast of Dorsetshire and Bath stone from Bath had become firm favorites (Section 8.7). Both are oolitic limestones; that is, stones formed by the agglomeration of tiny (less than 2 mm in diameter) spherical concretions of calcium carbonate, easily carved. Bath stone has a creamy yellow color, and

Portland stone is white. As London grew in the seventeenth and eighteenth centuries new squares were laid out in a uniform style and material. Portland or Bath stone were used for the best squares and the cheaper brick was employed elsewhere. In the eighteenth century a determined search was made for a form of construction that looked like stone but cost only as much as brick. The answer was stucco painted to resemble Bath stone. There was still the problem of durability, however. Inigo Jones had used stucco made with lime mortar on several of his buildings in the early seventeenth century, but it had not been a success and the practice lapsed until more durable materials were produced.

Adam used stucco—perhaps Liardet's patent—in Hanover Square about 1776; Wyatt covered the (still existing) front of 9 Conduit Street with Higgins' cement in 1779; Nash first used stucco in Bloomsbury Square in 1783, adopted Parker's Roman Cement on its introduction in 1796, and continued to use it till he changed to Hamelin's mastic or Dehl's mastic about 1819–1821. More important to us than this list of dates is the general truth that various patent stuccos came in just about the time of the Building Act of 1774, and were sparingly used until Nash inaugurated the real stucco age with the building of Park Crescent in Parker's Roman Cement in 1812 (Ref. 8.10, pp. 129–130).

In the eighteenth and early nineteenth centuries the facades were carefully scored with horizontal and vertical lines to represent joints in the masonry. Often each separate stone was painted to imitate the weathering of real stone. The now familiar plain paint was substituted much later, partly to save money and partly because the taste had changed.

Another eighteenth-century innovation was sculptured terracotta made to look like carved stone. The best-known type in London was Coade Stone, made from 1767 to 1835. It was widely used in private and public buildings, including Buckingham Palace and Westminster Abbey.

The eighteenth century was also in a sense an age of fake in design. The Florentine Renaissance had proclaimed its intention to revive Roman architecture, but in fact it produced an original style; the *palazzi* of Florence and Brunelleschi's dome made only a passing acknowledgment to ancient Rome, but the eighteenth century first consciously resurrected the style of Palladio and then revived Greek architecture.

It is curious that before the Greek Revival so little was known about Greek architecture. Books on architectural design were based, directly or indirectly, on what Vitruvius had said about his Greek sources. In the eighteenth century architects, archaeologists (a new breed), and *amateurs,* such as members of London's Dilettanti Society, set out for Greece, then under Turkish occupation, and for southern Italy and Sicily, where the great Doric temples at Paestum and Agrigentum were readily accessible for all to see but had been ignored as too primitive. The old buildings were carefully measured and the first accurate drawings of the Parthenon were published in 1789 in Volume 2 of *The Antiquities of Athens* (Ref. 8.12, p. 15).

The eighteenth and early nineteenth centuries were an era of great achievement in science and technology and also in the *history* of architecture, but their architecture was less original than that of the preceding four centuries. Nevertheless, the Palladian and the Greek Revival produced many elegant buildings, particularly in North America and in the newly founded colonies of Australia (Fig. 8.19), where architectural taste was lagging behind Europe and Greek columns were still fashionable in the second half of the nineteenth century.

8.19

Bligh House (43 Lower Fort Street, just below the Sydney Harbour Bridge),
built about 1833, now the Australian College of General Practitioners.

An interesting feature, both in America and Australia, is the construction of Greek
columns in timber, some of exquisite workmanship. This was an imitation in timber
of Greek stone columns which were themselves copied from timber columns in an
earlier age.

8.9   ENVIRONMENTAL ASPECTS

During the eighteenth century there were great improvements in the mechanics of
water supply. An increasing number of pumps operated by water power or, toward
the end of the century, steam were installed. Pipes were laid to deliver this water to
taps in public places and private houses. In fact, plumbing was taking on a modern
appearance (Fig. 8.20). The quality of the water, however, hardly improved, and this
problem remained unsolved until the second half of the nineteenth century. Similarly, there was little progress in sewage disposal. Indeed, in the early years of that
century there was considerable deterioration of hygiene in the fast-growing cities
(Section 8.1).

Significant advances were made in the art of heating in the late eighteenth and
early nineteenth centuries. Fireplaces were improved to ensure proper exhaustion of

8.20

Water pipes in Paris (from C. M. de la Gardette: *L'art du plombier et du fontainier*, Paris 1773).

smoke. In Adam's buildings fireplaces were always a prominent and much admired feature.

In the new factories waste heat from the steam engine operating the power plant was sometimes used, and in 1784 James Watt had his office heated by steam from his plant. Boulton and Watt thereafter provided steam heating in many of the factories they built.

In the first decade of the nineteenth century William Strutt built a hospital for Derby with subscriptions raised mainly from the local industrialists. He installed gravity warm-air heating and an exhaust system operated by weather vanes:

He built a noble building; hot air from below conveyed by a cockle all over the house. The whole institution a most noble and touching site (letter from the novelist Maria Edgeworth, quoted by Hacker, Ref. 8.11).

The Derbyshire General Infirmary was one of the first modern hospitals. It had a separate fever house which had no connection with the other wards, some smaller wards with two to three beds, and day rooms for convalescents. This was a considerable advance from the type of hospital that consisted of only one large room for the ill, the very ill, the dying, and both surgical and fever cases.

Shortly after Joseph Bramah used hot-water radiators to heat Westminster Hospital.

In 1734 a fan devised by Dr. Desaguliers was installed in the British Houses of Parliament to improve the ventilation. It consisted of a paddle wheel, 7 ft (2.1 m) in diameter, with 1-ft (300-mm) radial blades rotating in a concentric casing. It remained in use for eighty years.

Artificial lighting also improved during the eighteenth century. The new sperm-whale fisheries produced fuel to supplement the supplies of wax and tallow, and candlemaking was mechanized by the invention of machines that did not differ greatly from those used today. During the eighteenth century vast chandeliers with numerous candles became common among the well-to-do and balls continued well into the night.

The discovery of mineral oil supplemented the supply of lamp oil.

The coking of coal produced coal gas (Section 8.6) which was put to use for the first time in Whitehaven in northern England in 1765. In 1806 Boulton and Watt installed gas lighting for the east wing of Milford Old Mill (Ref. 8.5, p. 203).

The period 1700 to 1815 reviewed in this chapter was a great age for music. Its composers included Haydn, Mozart, and Beethoven. The science of acoustics, however, progressed only slowly.

Among those who contributed to its early theory were the French composer Jean-Phillippe Rameau and the Italian violinist Giuseppe Tartini.

The eighteenth and early nineteenth centuries marked the end of an era for most branches of building science. The great revolution that transformed the design of buildings came in the second half of the nineteenth and is considered in a separate book (Ref. 1.3).

REFERENCES

1.1 J. NEEDHAM: *Clerks and Craftsmen in China and the West.* Cambridge University Press, London, 1970. 470 pp.

1.2 *The Sketchbook of Villard de Honnecourt* (edited by T. BOWIE): Indiana University Press, Bloomington, 1966. 144 pp. A facsimile and translation of one of the few sketchbooks by a Gothic master mason, probably compiled between 1225 and 1250. Facsimile of a manuscript in the Bibliothèque Nationale, Paris.

1.3 H. J. COWAN: *Science and Building.* Wiley, New York, 1977, 368 pp.

2.1 C. FENSTERBUSCH (translator): *Vitruv—Zehn Bücher* über Architektur (Vitruvius—Ten Books on Architecture). Wissenschafliche Buchgesellschaft, Darmstadt, 1964. 585 pp.

2.2 R. ENGELBACH: *The Problem of the Obelisks from a Study of the Unfinished Obelisk at Aswan.* Unwin, London, 1923, pp. 17, 31.

2.3 MARCUS VITRUVIUS POLLIO (translated by M. MORGAN, Oxford, 1914): *The Ten Books of Architecture.* Dover, New York, 1960. 331 pp. This is a modern translation, based on the original first-century text, as far as that can be ascertained, and deleting modern additions.

2.4 LEONE BATTISTA ALBERTI (translated by J. LEONI): *Ten Books of Architecture.* Tiranti, London, 1955. 256 pp. This is a facsimile of the first English edition of 1755. Alberti reputedly presented the *Ten Books* to Pope Nicholas V in 1452. It was first printed in Latin in 1485 and in Italian in Venice in 1546. Leoni, a Venetian architect, used the Italian version.

2.5 SIR HENRY WOTTON: *The Elements of Architecture.* The University Press of Virginia, Charlottesville, 1968. lxxxv + 139 pp. This is a facsimile of the first edition published in London by John Bill, 1624.

2.6 R. J. FORBES and E. J. DIJKSTERHUIS: *A History of Science and Technology.* Penguin, Harmondsworth, 1963. Two volumes, 536 pp.

2.7 J. D. BERNAL: *Science in History.* Third Edition. Penguin, Harmondsworth, 1965. Four volumes, 1328 pp. Fourth Edition, 1969.

2.8 B. FARRINGTON: *Greek Science.* Penguin, Harmondsworth 1963. 320 pp.

2.9 M. R. COHEN and I. E. DRABKIN: *A Source Book in Greek Science.* Harvard University Press, Cambridge (Mass.), 1966. 581 pp. This contains translated passages from ancient Greek scientific texts.

2.10 O. NEUGEBAUER: *The Exact Science in Antiquity.* Brown University Press, Providence (R.I.), 1957. 240 pp.

2.11 W. W. ROUSE BALL: *A Short Account of the History of Mathematics.* Dover, New York, 1960. 522 pp. This is a reprint of the fourth edition of 1908; it excludes the twentieth century.

2.12 C. B. BOYER: *A History of Mathematics.* Wiley, New York, 1968. 717 pp.

2.13 G. GUARINI: *Architettura Civile.* Edizioni Polifilio, Milan, 1968. 471 pp. A reprint of the work published by Guarini in 1683, describing his own buildings.

2.14 A. F. BURSTALL: *Simple Working Models of Historic Machines, Easily Made by the Reader.* Arnold, London, 1968. 79 pp.

2.15 A. G. KELLER: *A Theatre of Machines.* Chapman and Hall, London, 1964. 115 pp.

3.1 A. BADAWY: *Architecture in Ancient Egypt and the Near East.* M.I.T. Press, Cambridge (Mass.), 1967. 246 pp.

3.2   H. F. MUSSCHE: *Monumenta Graeca et Romana*. Volume 2, Greek Architecture, Part 1, Religious Architecture, J. E. Brill, Leiden (Holland), 1968. Plates 11 and 12.

3.3   E. GOSE: *Die Porta Nigra in Trier* (The Porta Nigra—The Black Gate in the German city of Trier). Mann, Berlin, 1969. Two volumes. 172 pp. + 268 plates.

3.4   B. CUNLIFFE: *Roman Bath Discovered.* Routledge and Kegan Paul, London, 1971. 108 pp.

3.5   I. A. RICHMOND: Roman Timber Building, in *Studies in Building History,* Ref. 6.26, pp. 15–26.

3.6   E. GIBBON: *The Decline and Fall of the Roman Empire.* An Abridgement by D. M. LOW. The Reprint Society, London, 1960. 924 pp.

3.7   M. GOLDFINGER: *Villages in the Sun.* Lund Humphries, London, 1969. 224 pp.

3.8   H. KAHLER: *Der Römische Tempel* (The Roman Temple). Mann, Berlin, 1970. 42 pp. + 72 plates.

3.9   H. J. COWAN: *Architectural Structures.* Elsevier, New York, 1971. 400 pp.

3.10   W. B. DINSMOOR: *The Architecture of Ancient Greece.* Third Edition. Norton, New York, 1975. 424 pp.

3.11   C. ROETTER: *Fire is their Enemy.* Angus and Robertson, Sydney, 1962. 184 pp.

3.12   G. J. VAROUFAKIS: Technical specifications of the 4th century B.C. *Technika Chronika,* Vol. 43, No. 2 (February 1974), pp. 155–162 (in Greek, with English summary). Includes the text of an inscription in the Museum of Eleusis which specifies the composition of a copper-tin alloy.

3.13   C. STANLEY: A brief history of concrete. *Building Technology and Management,* Vol. 12 (October 1974), pp. 5–6.

3.14   A. JAENICKE and H. GOETZ: *Mamallapuram.* Scherpe, Krefeld (West Germany), 1966. 16 pp. + 85 plates.

3.15   ROWLAND J. MAINSTONE: *Developments in Structural Form.* Allen Lane, London, 1975. 350 pp.

3.16   SOMERS CLARKE and R. ENGELBACH: *Ancient Egyptian Masonry—The Building Craft.* Oxford University Press, London, 1930. 242 pp.

3.17   B. ALLSOPP: *A History of Classical Architecture.* Pitman, London, 1965. 215 pp.

3.18   S. GIEDION: *The Eternal Present: The Beginnings of Architecture.* Oxford University Press, London, 1964. 583 pp.

3.19   *World Architecture.* Paul Hamlyn, London, 1963. 348 pp.

3.20   BANISTER FLETCHER: *A History of Architecture by the Comparative Method.* Seventeenth Edition. Athlone, London, 1961. 1366 pp. (The eighteenth edition was published in 1975.)

3.21   G. DANIEL: *The First Civilizations.* Penguin. Harmondsworth, 1971. 201 pp.

3.22   R. CARPENTER: *The Architects of the Parthenon.* Penguin, Harmondsworth, 1970. 193 pp.

3.23   D. R. ROBERTSON: *Greek and Roman Architecture.* First Paperback Edition. Cambridge University Press, London, 1969. 407 pp.

3.24   J. HEYMAN: "Gothic" construction in Ancient Greece. *Journal of the Society of Architectural Historians,* Vol. 31 (1972), pp. 3–9.

3.25   S. B. HAMILTON: The structural use of iron in antiquity. *Trans. Newcomen Soc.,* Vol. 31 (1957–1959), pp. 29–47.

3.26   ANDREA PALLADIO: *The Four Books of Architecture.* Dover, New York, 1965. 110 pp. + 94 plates. A facsimile of the English edition published by Isaac Ware in London in 1738. It was first published in Italian in 1570.

3.27   G. E. RICKMAN: *Roman Granaries and Store Buildings.* Cambridge University Press, London, 1971. 349 pp.

3.28   R. MERRIFIELD: *Roman London.* Cassell, London, 1969. 212 pp.

3.29   H. PLOMMER: *Vitruvius and Later Roman Building Manuals.* Cambridge University Press, London, 1973. 117 pp. This includes the full Latin text and English translation of the treatise *De diversis fabricis architectonicae,* by Cetius Faventinus.

3.30   MARION ELIZABETH BLAKE: *Ancient Roman Construction in Italy from the Prehistoric Period to Augustus.* Carnegie Institute, Publication 570, Washington, D.C., 1947. 421 pp. + 57 plates.

3.31   M. E. BLAKE: *Roman Construction in Italy from Tiberius through to Flavians.* Carnegie Institute, Publication 616, Washington, D.C., 1959. 195 pp. + 31 plates.

3.32 M. E. BLAKE: *Roman Construction in Italy from Nerva through the Antonines*. American Philosophical Society, Memoirs, Volume 96, Philadelphia, 1973. 304 pp. + 36 plates.

3.33 N. DAVEY: Roman concrete and mortar. *Structural Engineer,* Vol. 52 (1974), pp. 193–195.

3.34 G. HAEGERMANN, G. HUBERTI, and H. MOLL: *Vom Caementum zum Spannbeton.* (From Caementum to Prestressed Concrete.) Bauverlag, Wiesbaden, 1964. Two volumes, 491 pp.

3.35 L. SPRAGUE DE CAMP: *The Ancient Engineers*. Rigby, Adelaide, 1963. 408 pp. Original edition published in America by Doubleday in 1960.

3.36 H. STRAUB: *A History of Civil Engineering.* Leonard Hill, London, 1952. 258 pp.

4.1 P. G. ISAAC: *Roman Public-Works Engineering.* Bulletin No. 13, Department of Civil Engineering, King's College, Newcastle, 1958. 16 pp. Mr. Isaac later became Professor of Public Health Engineering at the University of Newcastle.

4.2 L. L. BERANEK: *Acoustics, Music and Architecture.* Wiley, New York, 1962. 586 pp.

4.3 L. WRIGHT: *Clean and Decent—The Fascinating History of the Bathroom and the W.C.* Routledge and Kegan Paul, London, 1960. 282 pp.

4.4 W. M. WINSLOW: *A Libation to the Gods.* Hodder and Stoughton, London, 1963. 191 pp.

4.5 J. CARCOPINO: *Daily Life in Ancient Rome.* Penguin, Harmondsworth, 1956, p. 49.

4.6 *Encyclopaedia Britannica.* Ninth Edition. Adam and Charles Black, Edinburgh, 1875–1888. Twenty-five volumes.

4.7 E. G. RICHARDSON: *Acoustics for Architects.* Arnold, London, 1945, pp. 19–26.

4.8 N. DAVEY. *A History of Building Materials.* Phoenix House, London, 1961. 260 pp.

5.1 J. H. de WALLER and A. C. ASTON: Corrugated Shell Roofs. *Proc. Inst. Civ. Eng.,* Vol. 2 (1953), pp. 153–196.

5.2 G. RIVANI: *Le Torre di Bologna* (The Towers of Bologna). Tamari, Bologna, 1966. 251 pp.

5.3 D. H. S. CRANAGE: *Cathedrals and How they Were Built.* Cambridge University Press, London, 1951. 42 pp. + 20 plates.

5.4 K. TANGE and N. KAWAZOE: *Ise.* M.I.T. Press, Cambridge (Mass.), 1965. 212 pp.

5.5 J. NEWLANDS: *The Carpenter's and Joiner's Assistant.* Blackie, London, undated (probably late nineteenth century). 254 pp. + 100 plates.

5.6 J. FITCHEN: *The Construction of Gothic Cathedrals.* Oxford University Press, London, 1961. 344 pp.

5.7 E. BALDWIN SMITH: *The Dome—a Study in the History of Ideas.* Princeton University Press, Princeton (N.J.), 1971. 164 pp.

5.8 HASSAN FATHY: *Architecture for the Poor.* Chicago University Press, Chicago, 1973. 233 pp.

5.9 C. A. HEWETT: *English Cathedral Carpentry.* Wayland, London, 1974. 176 pp.

5.10 PROCOPIUS: *Buildings* (Greek text with an English translation by H. B. DEWING). Heinemann, London, 1954. 542 pp.

5.11 R. J. MAINSTONE: The structure of the church of St. Sophia, Istanbul. *Transactions of the Newcomen Society,* Vol. 38 (1965–1966), pp. 23–49.

5.12 JANE A. WIGHT: *Brick Building in England from the Middle Ages to 1550.* John Baker, London, 1972. 439 pp.

5.13 J. HEYMAN: Beauvais Cathedral. *Trans. Newcomen Soc.,* Vol. 40 (1967–1968), pp. 15–35.

5.14 D. B. HARDEN: Domestic window glass—Roman, Saxon, Medieval, in *Studies in Building History* Ref. 6.26, pp. 39–63.

5.15 W. T. O'DEA: *A Short History of Lighting.* H. M. Stationery Office, London, 1958. 40 pp.

5.16 J. HEYMAN: On shell solutions of masonry domes. *International Journal of Solids and Structures,* Vol. 3 (1967), pp. 227–241.

5.17 F. W. B. CHARLES: *Medieval Cruck-Building and Its Derivatives.* Society for Medieval Archaeology, London, 1967. 70 pp. + 32 plates.

5.18 J. HEYMAN: Westminster Hall roof. *Proc. Inst. Civ. Eng.,* Vol. 37 (1967), pp. 137–162.

5.19 H. JANTZEN: *Die Hagia Sophia des Kaiser Justinian in Konstantinopel* (The Hagia Sophia of the Emperor Justinian in Constantinople). Schauberg, Cologne, 1967. 109 pp.

5.20 B. UNSAL: *Turkish Islamic Architecture, Seljuk to Ottoman.* Alec Tiranti, London, 1970. 116 pp. + 130 plates.

5.21 A. U. POPE: *Persian Architecture.* Thames and Hudson, London, 1965. 287 pp.

6.1 J. HARVEY: *The Master Builders—Architecture in the Middle Ages.* Thames and Hudson, London, 1971. 144 pp.

6.2 G. G. UNGEWITTER: *Lehrbuch der gothischen Konstruktionen* (Manual of Gothic Construction). Fourth edition. Tauchnitz, Leipzig, 1901. Two volumes.

6.3 P. FRANKL: *The Gothic-Literary Sources and Interpretations through Eight Centuries.* Princeton University Press, Princeton (N.J.), 1960. 916 pp.

6.4 R. J. MAINSTONE: Structural theory and design. *Architecture and Building,* Vol. 43 (1959), pp. 106–113, 186–195, and 214–221.

6.5 O. von SIMSON: *The Gothic Cathedral—Origins of Gothic Architecture and the Medieval Concept of Order.* Harper and Row, New York, 1964. 275 pp.

6.6 Sir G. G. SCOTT: *Lectures on the Rise and Development of Medieval Architecture Delivered at the Royal Academy.* London, 1879. Lecture XV, pp. 212–213.

6.7 E. E. VIOLLET-LE-DUC: *Dictionnaire Raisonné de l'Architecture du XIe à XVIe Siècle.* Librairies-Impremieries Reunies, Paris, 1858–1868. Ten volumes.

6.8 J. F. FITCHEN: The erection of French Gothic nave vaults. *Gazette des Beaux Arts,* Vol. 55 (1960), pp. 281–300.

6.9 W. WHEWELL: Description of a mode of erecting light vaults over churches and similar spaces. *Journal of the Royal Institution of Great Britain,* 1831. Quoted by Fitchen, Ref. 5.6, pp. 180–181.

6.10 A. J. S. PIPPARD and J. F. BAKER: *The Analysis of Engineering Structures,* Fourth Edition. Arnold, London, 1968, pp. 391–394.

6.11 J. HEYMAN: On the rubber vaults of the Middle Ages, and other matters. *Gazette des Beaux-Arts,* Vol. 63 (1968), pp. 177–188.

6.12 H. MOSELEY: *The Mechanical Principles of Engineering and Architecture.* London, 1843. Quoted by Heyman (Ref. 6.11).

6.13 W. J. M. RANKINE: *Manual of Applied Mechanics.* 1858. Quoted by Timoshenko (Ref. 8.16).

6.14 P. ABRAHAM: *Viollet-le-Duc et le Rationalisme Médiéval.* Fréal, Paris, 1934. 116 pp.

6.15a. R. MARK: The structural analysis of Gothic cathedrals. *Scientific American,* November 1972, pp. 90–99.
   b. R. MARK, J. F. ABEL and K. O'NEILL: Photoelastic and finite-element analysis of a quadripartite vault. *Experimental Mechanics,* Vol. 13 (1973), pp. 322–329.
   c. R. MARK and M. I. WOLFE: The Collapse of the Beauvais Vaults of 1284. *Working Paper No. 14. School of Architecture and Urban Planning,* Princeton University, Princeton (N.J.), 1975. 23 pp.

6.16 E. FUMAGALLI: *Statical and Geomechanical Models.* Springer, Vienna, 1973. 182 pp.

6.17 E. VIOLLET-LE-DUC: *Lectures on Architecture.* Simpson, Low, Marston, Searle, and Rivington, London, 1881. Volume 2, 468 pp.

6.18 F. E. HOWARD: Fan vaults. *Archeol. J.,* Vol. 68, No. 1 (1911), quoted by Heyman, Ref. 6.28.

6.19 H. R. HAHNLOSER: *Villard de Honnecourt—Kritische Ausgabe des Bauhüttenbuches MS FR 19093 der Pariser Nationalbibliothek.* (Villard de Honnecourt. Critical Complete Edition of the Stone Mason's Lodge Book MS Fr 19093 of the Paris National Library.) Second edition. Akademische Druck and Verlagsanstalt, Graz (Austria), 1972. 403 pp. + 96 plates. This treatise contains a facsimile of Villard's notebook, the text in Old French and in German translation, and a commentary more than ten times the length of the original booklet.

6.20 F. MACKENZIE: *Observations on the Construction of the Roof of King's College Chapel, Cambridge.* John Weale, London, 1840. 20 pp.

6.21 G. ROSENBERG: The functional aspect of the Gothic style. *Jnl. Roy. Inst. Brit. Arch.,* Vol. 43 (1936), pp. 273–290.

6.22   F. B. ANDREWS: *The Medieval Builder and His Methods*. Oxford University Press, Oxford, 1926. 99 pp.

6.23   R. J. MAINSTONE: Structural theory and design before 1742. *Architectural Review*, Vol. 113, No. 854 (April 1968), pp. 303–310.

6.24   RICHARD BROWN: *Sacred Architecture: Its Rise, Progress and Present State*, Fisher, London, 1845. Quoted by Fitchen, Ref. 5.6, p. 275.

6.25   JOHN RUSKIN: *The Stones of Venice*. Edited and abridged by J. G. LINKS. Collins, London, 1960. 254 pp. This is a modern abridgement of the book first published in 1853.

6.26   E. M. JOPE (Ed.): *Studies in Building History*. Odhams, London, 1961. 287 pp. Fourteen essays on building from Roman Britain to seventeenth century Ireland.

6.27   J. HEYMAN: The stone skeleton. *Int. J. Solids Struct.*, Vol. 2 (1966), pp. 249–279.

6.28   J. HEYMAN: Spires and fan vaults. *Int. J. Solids Struct.*, Vol. 3 (1967), pp. 243–257.

6.29   BERNARD FEILDEN: Private communication.

6.30   D. J. DOWRICK and P. BECKMANN: York Minster structural restoration. *Proc. Inst. Civil Engineers*, Vol. 49 (1971), Paper 7415 S, pp. 93–156. A graphical analysis of the thrust line for the Central Tower, made in 1970 by Ove Arup and Partners.

7.1   KENNETH K. CLARK: *Civilization*. Murray, London, 1969. 359 pp.

7.2   *The Shorter Oxford English Dictionary*. Oxford University Press, Third Edition. London, 1956. *Dome*, p. 550.

7.3   Publications of the Wren Society.
       a.   Volume XIII. 205 pp. + 36 plates.
       b.   Volume XV. 232 pp. + 104 plates.
       c.   Volume XVI. 216 pp. + 22 plates.

7.4   H. J. HOPKINS: *A Span of Bridges*. David and Charles, Newton Abbott (England), 1970. 288 pp.

7.5   EDWARD MacCURDY: *The Notebooks of Leonardo da Vinci*. Two volumes. The Reprint Society, London, 1954. 509 + 566 pp.

7.6   ARTHUR MORLEY: *Strength of Materials*. Eighth Edition. Longmans, London 1934. 569 pp.

7.7   L. L. B. CARNEIRO: Galileo, founder of the science of the strength of materials. *RILEM Bulletin*, No. 27 (June 1965), pp. 100–119.

7.8   LAWRENCE WRIGHT: *Clean and Decent*. Routledge and Kegan Paul, London, 1960. 282 pp.

7.9   ALLARDYCE NICOLL: *The Development of the Theatre*. Fifth Edition. Harrap, London, 1966. 292 pp.

7.10   JOAN GADOL: *Leon Battista Alberti—Universal Man of the Early Renaissance*. University of Chicago Press, Chicago, 1969. 266 pp.

7.11   *Filarete's Treatise on Architecture* (edited by J. R. SPENCER). Yale University Press, New Haven, 1966. Two volumes, 339 pp. + facsimile of the manuscript. The first original book on the architecture of the Renaissance.

7.12   R. J. MAINSTONE: Brunelleschi's dome S. Maria dei Fiore and some related structures. *Trans. Newcomen Soc.*, Vol. 42 (1969–1970), pp. 107–126.

7.13   H. KLOTZ: *Die Frühwerke Brunelleschi's und die mittelalterliche Tradition* (The Early Work of Brunelleschi and the Medieval Tradition). Gebr. Mann Verlag, West Berlin, 1970. 149 pp. + 241 plates.

7.14   F. PRAGER and GUSTINA SCAGLIA: *Brunelleschi*. M.I.T. Press, Cambridge (Mass.), 1970. 152 pp.

7.15   W. B. PARSONS: *Engineers and Engineering in the Renaissance*. M.I.T. Press, Cambridge (Mass.), 1967. 661 pp. Originally published, posthumously, in 1939 bv Williams and Wilkins.

7.16   J. HEYMAN: *Coulomb's Memoir on Statics*. Cambridge University Press, London, 1972. 212 pp. This book contains a facsimile of the *Essai sur une application des règles de Maximis & Minimis à quelques Problèmes de Statique, relatifs a l'Architecture*, originally published in the *Memoires de Mathématique & de Physique, présentés a l'Academie Royale des Sciences par divers Savants, & lus dans ses Assemblées*, Vol. 7 (1773), pp. 343–384, printed in Paris in 1776. Includes an English translation and a detailed explanatory commentary by Professor Heyman. The memoir contains many important solutions, including Coulomb's (correct) solution of the theory of bending.

7.17 J. SUMMERSON: *Sir Christopher Wren.* Collins, London, 1953. 160 pp.

7.18 C. TRUESDELL: *Essays in the History of Mechanics.* Springer, New York, 1969. 384 pp.

7.19 GALILEO GALILEI: *Discorsi e Dimostrazioni Matematiche Intorno a Due Nuove Scienze Attenenti all Mecanica & i Movimento Locali.* Paolo Boringhieri, Turin, 1958. 887 pp. This is a reprint (but not a facsimile) in Italian only, of the book originally published by Elzevir in Leiden in 1638. It includes Galileo's theory of bending and his discourse on the strength of materials.

7.20 J. S. ACKERMAN: *The Architecture of Michelangelo.* Penguin, London, 1970. 373 pp.

7.21 E. RODENWALDT: *Leon Battista Alberti—ein Hygieniker der Renaissance.* Springer, Heidelberg, 1968. 104 pp.

7.22 N. PEVSNER: *An Outline of European Architecture.* Seventh Edition. Penguin, Harmondsworth, 1972. 446 pp.

7.23 L. L. BERANEK: *Music, Acoustics, and Architecture.* Wiley, New York, 1962. 586 pp.

7.24 W. C. SABINE: *Collected Papers on Acoustics.* Dover, New York, 1964, 279 pp.

7.25 P. H. SCHOFIELD: *The Theory of Proportion in Architecture.* Cambridge University Press, London, 1958. 156 pp.

7.26 R. WITTKOWER: *Architectural Principles in the Age of Humanism.* Third Edition. Alec Tiranti, London, 1962. 173 pp. + 48 plates.

7.27 *Lives of the Engineers, Selections from Samuel Smiles* (edited by T. P. HUGHES). M.I.T. Press, Cambridge (Mass.), 1966. 447 pp.

7.28 MARIANO DI JACOPO DETTO IL TACCOLA: *Liber Tertius de Ingeneis ac Edifitiis non Usitatis.* Edizioni il Polifolio, Milan, 1969. 156 pp. Facsimile of a fifteenth-century manuscript, with printed version of the handwritten Latin text.

7.29 N. BILLINGTON: A historical review of the art of heating and ventilating. *Architectural Science Review,* Vol. 2 (1959), pp. 118–130.

7.30 E. G. RICHARDSON: Acoustics in modern life. *Science Progress* Vol. 42 (1954), pp. 232–239.

7.31 R. McGRATH et al.: *Glass in Architecture and Decoration.* Architectural Press, London, 1961. 712 pp.

8.1 GEORGE GODWIN: *Town Swamps and Social Bridges.* Leicester University Press, Leicester, 1972, 102 pp. (Originally published in 1859.)

8.2 R. MAGUIRE and P. MATTHEWS: The Iron Bridge at Coalbrookdale. *Architectural Association Journal,* Vol. 74 (August 1958), pp. 30–45.

8.3 J. A. WILLIAMS: Remedial Works for the Iron Bridge. *Ground Engineering,* May 1974, pp. 36–41.

8.4 DAVID B. STEINMAN and SARA RUTH WATKINS: *Bridges and Their Builders.* Dover, New York, 1957, 401 pp.

8.5 H. R. JOHNSON and A. W. SKEMPTON: William Strutt's cotton mills. *Trans. Newcomen Soc.,* Vol. 30 (1955–1957), pp. 179–205.

8.6 A. W. SKEMPTON and H. R. JOHNSON: The first iron frames. *Architectural Review,* Vol. 114 (March 1962), pp. 175–186.

8.7 R. J. BARFOOT: The Aspdin jigsaw. *Concrete* (London), Vol. 8 (August 1974), pp. 18–26.

8.8 G. M. A. RICHTER: *The Furniture of the Greeks, Etruscans and Romans.* Phaidon, London, 1966. 143 pp.

8.9 HERBERT CESCINSKY: *English Furniture from Gothic to Sheraton.* Dover, New York, 1968. 406 pp.

8.10 JOHN SUMMERSON: *Georgian London.* Barrie and Jenkins, London, 1962. 349 pp.

8.11 L. C. HACKER: William Strutt of Derby. *Jnl. of the Derbyshire Archaeological and Natural History Society,* Vol. 80 (1960), pp. 49–70.

8.12 J. MORDAUNT CROOK: *The Greek Revival.* John Murray, London, 1972. 204 pp.

8.13 M. KRANZERG and C. W. PURSELL (Eds.): *Technology in Western Civilization.* Oxford University Press, New York, 1967. Volume I, 802 pp.

8.14 JOHN SMEATON: *The Narrative of the Building of the Edystone Lighthouse with Stone*. Longman, Hurst, Rees, Orme and Brown, London, 1813. 198 pp.

8.15 I. TODHUNTER and K. PEARSON: *A History of the Theory of Elasticity*. Dover, New York, 1960. 3 volumes, 2244 pp. A reprint of the 1886–1893 edition.

8.16 S. P. TIMOSHENKO: *History of Strength of Materials*. McGraw-Hill, New York, 1953. 452 pp.

8.17 T. TREDGOLD: *Practical Essay on the Strength of Cast Iron*. J. Taylor, London, 1824. 305 pp. A standard textbook in the early nineteenth century, still available in many libraries.

8.18 S. GIEDION: *Space, Time and Architecture*. Fifth Edition. Harvard University Press, Cambridge (Mass.), 1967. 897 pp.

GLOSSARY

This glossary of technical terms used in the preceding chapters is intended to assist the general reader. Words in italics denote a cross-reference.

aqueduct   A channel or conduit especially the part carried above ground on arches for the conveyance of water.

arch   A structure that supports a load across a horizontal opening mainly by compression. See also *corbeled masonry arch* and *true masonry arch*.

architrave   The lowest of the three parts of the entablature of a Greek temple. It lies beneath the frieze and rests on the capital of the column. Hence it often acts like a *beam*.

beam   A structural member that supports a load across a horizontal opening by bending.

bending moment   Moment in a member of a structure caused by the loads acting on the structure.

brittle material   A material that breaks with little or no *plastic deformation*. Stone, brick, cast iron, glass, and concrete are brittle materials.

buckling   Failure of a compression member by deflection at right angles to the load. The material is not necessarily damaged by buckling.

built-in   Rigidly restrained at the ends to prevent rotation.

buttress   A projecting structure built against a wall to resist a horizontal force. A flying buttress is a strut or arch rising from a pier and abutting the wall.

camber   A slight upward curvature given to a beam, girder, or truss to compensate for its anticipated deflection.

came   Lead strip of H-section used to hold pieces of glass in a window.

cantilever   A projecting beam, slab, truss, or column *built-in* at one end and free at the other.

cast iron   Iron with a total carbon content between 1.8 and 4.5%. It is hard and strong in compression but also a *brittle material* and weak in tension.

catenary   Curve assumed by a cable hanging under its own weight; that is, the cable is purely in tension. A similar structure turned upside down forms a catenary arch that is purely in compression.

chord   A horizontal member in a *truss*.

concrete   An artificial rock usually made from pieces of stone or gravel, sand, and a binding material, such as lime or cement.

corbel   A stone or brick laid horizontally and projecting from the surface of a wall; it is, in fact, a short *cantilever*.

corbeled masonry arch   An arch formed by a series of corbels that gradually close the opening. See also *true masonry arch*.

crown   The highest point of an *arch* or *dome*. The stone at the crown of a masonry arch is called the *keystone*.

cupola   A dome.

cylindrical vault   A vault formed by a portion of a hollow cylinder, most commonly half a circular cylinder; that is, any cross section of the vault is a semicircle.

deflection   Deformation due to bending.

diorite   A hard, igneous rock.

dome   A vault of double curvature formed by the rotation of a curve around a vertical axis. The most common type is the spherical dome formed by the rotation of a part of a circle. See also *hemispherical dome* and *shallow spherical dome*.

drum   A vertical wall in the shape of a hollow cylinder supporting a dome.

elastic deformation   Deformation fully recovered when the load is removed.

entasis   An almost imperceptible swelling given to Greek and later Classic columns to correct the optical illusion of concavity which would result if the sides were actually straight.

factor of safety   The ratio of the stress at failure to the maximum permissible or working stress.

flexure   Bending.

flying buttress   See *buttress*.

formwork   Temporary structure used during construction, particularly for supporting wet concrete to which it gives its form.

giga   Prefix meaning a thousand million; a gigapascal is thus 1000 megapascal.

girder   A large *beam*. A lattice girder is a *truss*.

Golden Section   A geometric construction for dividing a line into two unequal parts *a* and *b,* such that $b/a = \frac{1}{2}(1 + \sqrt{5}) = 1.618$.

gypsum   Calcium sulfate dihydrate ($CaSO_4 \cdot 2H_2O$), a natural mineral which is the raw material of gypsum plaster. The ancient Egyptians used gypsum as a mortar.

hammer-beam roof   A medieval timber roof without a *tie*. The hammer beams are supported on brackets on the walls, but unlike the tie of a modern roof truss they do not meet.

harmonic proportions   The proportions that produce simple harmonies in music. At various times these proportions, particularly $5/3 = 1.667$, were held to produce harmony in architectural design.

header   A brick (or block or stone) laid across the wall to bond the *stretcher* bricks.

hemispherical dome   A *dome* formed by the rotation of a semicircle. Its support *reactions* are vertical only. See also *shallow spherical dome*.

herringbone bond   Bricks or stone laid in zigzag pattern.

Hooke's Law   Stress is directly proportional to *strain*.

hoop force   Internal horizontal force in a dome.

hoop tension   The tension that occurs in the lower portion of a *hemispherical dome*.

horizontal reaction   A reaction that acts horizontally or the horizontal component of a support *reaction* inclined to the vertical. It must be resisted by a *tie* or *buttresses*.

hydraulic mortar   A mortar that is not washed out by water.

hypotenuse   The longest side of a right-angled triangle opposite the right angle.

incommensurable ratio   A ratio that cannot be expressed as a fraction of two *rational numbers*.

irrational number   A real number that is not *rational;* for example π, $\sqrt{2}$, and $\sqrt{5}$.

jack arch   Short-span arch that supports a floor between closely spaced beams.

keystone   The stone at the top of a masonry arch.

kilo   Prefix meaning a thousand.

lantern   Small open or glazed structure crowning a roof, particularly a dome.

limit design   Design based on the limiting loads at which a structure collapses because of the formation of open joints.

lintel   A short-span beam, usually over a door or window.

mega   Prefix meaning a million.

membrane theory   A theory for designing a thin structural shell on the assumption that it is free from bending.

milli   Prefix meaning one-thousandth.

modulus of elasticity   Measure of elastic deformation, defined as the *stress* that would produce a unit *strain*.

moment   A force multiplied by the distance at which it acts.

moment of resistance   Internal moment in a beam which for equilibrium must equal the *bending moment* acting on a beam.

monolithic   Cast in one piece and therefore continuous.

mortise-and-tenon joint   Traditional timber joint formed by a rectangular slot (mortise) into which a tongue (tenon) from another piece fits.

nave   The body of a church or cathedral, usually separated from the aisles by lines of pillars.

neutral axis   Line at which the flexural stresses change from tension to compression.

newton   The unit of force in the SI metric system, abbreviated N; 1 newton is the force that, applied to a mass of 1 kilogram, produces an acceleration of 1 meter per second per second.

oculus   A round window, particularly at the crown of a dome.

party wall   A wall that forms part of two buildings.

pascal   The unit of stress in the SI metric system, abbreviated Pa. It is defined as 1 *newton* per square meter.

pinnacle   A vertical pointed structure rising above a roof or *buttress*.

plain concrete   Concrete without reinforcement.

plastic deformation   Deformation not recovered when the load is removed.

pumice   A very light porous rock of volcanic origin.

Pythagorean triangle   A right-angled triangle whose sides are in the proportion 3:4:5. The enclosed angles are 90°, 53°7', and 36°53'.

rafter   A sloping piece of timber extending from the wall plate (i.e., a timber plate on top of a masonry wall) to the ridge or the sloping upper member of a roof truss.

rational number   An integer (whole number) or a fraction. See also *incommensurable ratio* and *irrational number.*

reaction   Force exerted by the ground or by another structural member in opposition to the loads.

resistance moment   See *moment of resistance.*

shallow spherical dome   A dome formed by the rotation of a circular arc which is less than a semicircle. Its support *reactions* are inclined. If the angle subtended by the circular arc at its center of curvature is less than 104°, the *hoop forces* are entirely compressive. See also *horizontal reaction.*

shell   Thin curved structural surface.

span   The distance between the supports of a structure.

spherical dome   See *hemispherical dome* and *shallow spherical dome.*

springings   Supports of an *arch* or *dome.*

statically determinate   Soluble by *statics* alone.

statically indeterminate   Insoluble by statics alone because there are more members, rigid joints, or *reactions* than there are statical equations.

statics   The branch of physics that deals with forces in equilibrium.

steel   An alloy of iron with a carbon content between 0.1 and 1.78%. Iron with a lower carbon content is *wrought iron;* iron with a higher carbon content is *cast iron.*

strain   Deformation per unit length.

stress   Force per unit area.

stress-strain diagram   The diagram obtained by plotting the stresses in a test piece against the strains. It is used to assess the structural suitability of materials.

stretcher   A brick (block or stone) laid with its length parallel to the wall. Stretchers are often interspersed with *headers* to achieve a proper bond.

strut   A compression member, the opposite of a *tie.*

tepidarium   A warm but not hot room in a Roman bath.

terracotta   Burnt clay units of a shape more complex than a brick or a tile.

thrust   Compressive force or *reaction.*

tie   A tension member, the opposite of a *strut.*

transept   The transverse part of a cruciform building at right angles to the *nave.*

true masonry arch   An arch whose *voussoirs* are arranged so that the joints are at right angles to the line of *thrust,* as opposed to a *corbeled masonry arch.*

truss   An assembly of straight tension and compression members which performs the same function as a deep *beam.*

tufa   A porous rock of volcanic origin.

ultimate load   The highest load that a structure can sustain.

vault   See *cylindrical vault* and *dome*.

vertical reaction   A *reaction* that acts vertically or the vertical component of an inclined reaction. See also *horizontal reaction*.

voussoir   Wedge-shaped block of masonry forming part of an *arch* or *dome*.

water-cement ratio   The ratio of water to cement in a concrete mix.

wrought iron   Iron with a carbon content lower than steel and consequently weaker. It was used as a structural material in the eighteenth and nineteenth centuries. It is not so strong in compression as cast iron but is more ductile and has a higher tensile strength.

Young's modulus   The *modulus of elasticity* in tension and in compression.

zenith   The point in the sky immediately above the observer; that is, at altitude 90°.

A NOTE ON UNITS OF MEASUREMENT

Numerical data are given in SI metric units and in American units (formerly used also in British Commonwealth countries). In the text the unit quoted from the original source is generally given first, with the conversion in brackets. The following conversion table lists units used more than once in this book, that is, units of length and of stress.

LENGTH

kilometers (km), meters (m), millimeters (mm)
miles, feet (ft), inches (in)

1 km = 1000 m, 1 m = 1000 mm   1 mile = 5280 ft, 1 ft = 12 in.
1 km = 0.621 371 miles   1 mile = 1.690 34 km
1 m = 3.280 ft = 3 ft 3½ in.   1 ft   = 0.304 80 m
1 mm = 0.039 37 in.   1 in. = 25.400 mm

STRESS

gigapascals (GPa), megapascals (MPa), kilopascals (kPa), pascals (Pa)
pounds per square foot (psf or lb per sq ft)
kilo pounds per square inch (ksi), pounds per square inch (psi or lb per sq in)

1 GPa = 1000 MPa, 1MPa = 1000 kPa, 1 kPa = 1000 Pa
1 MPa = 1 newton per square millimeter (N/mm²)
1 Pa = 1 newton per square meter (N/m²)
1 ksi = 1000 psi, 1 psi = 144 psf
1 GPa = 145.038 ksi   1 ksi = 6.894 76 MPa
1 MPa = 145.038 psi   1 psi = 6.894 76 kPa
1 kPa =   0.145 038 psi = 20.885 psf   1 psf = 47.880 3 Pa

PEOPLE, PLACES, AND STRUCTURES

(Titles of books, societies and institutions appear in the General Index)
Page numbers in italics refer to illustrations

People, Places, and Structures

Davey, N., 271
Davy, Humphry, 222
de La Hire, Philippe, 195, *195,* 225
della Porta, Giacomo, 192, 193
Delphi, Theban Treasury, 44
Denison, Edmund Beckett (Lord Grimthorpe),
    201, 202
Derbyshire General Infirmary, 267
Diana, Temple of, Ephesus, 15, 27
Diderot, Denis, 221
Dinsmoor, W.B., 45, 270
Diocletian, Baths of, Rome, 129, 135
    Palace of, Spalatum, 72, 129
Dionyssos, Theater of, Athens, 85, *85*
Dome of the Rock, Jerusalem, 103, *105, 106,* 129
Domus Aurea (Golden House), Rome, 57, 63, 70
Duomo of Florence, 7, 100, 170, 171, *175,* 182,
    198, 205
Durham Cathedral, 129, 136, 142

Eddystone Lighthouse, 256, *257, 258,* 275
Egypt, architecture, 27, 269
Elgin, Lord, 46
Ely Cathedral, 152, *153,* 154
Engelbach, R., *38,* 269, 270
Engels, Friedrich, 239
Ensingen, Ulrich von, 137
Epidauros, Theater at, 85
Euclid, 11, 14, 16, 108, 141
Eudoxus, 16
Euler, Leonard, 227, 234
Evelyn, John, 182

Farnese Palace, Rome, *211*
    Theater, Parma, 211
Faventinus, 11, 12, 55, 56, 84, 92
Felice aqueduct, Rome, 207
Fensterbusch, C., 12, 269
Filarete, 169, 170, 273
Fioravante, Neri di, 171
Fitchen, J.F., 142, *143, 146,* 161, 173, 271, 272
Flavian Amphitheater, *see* Colosseum
Fletcher, Banister, 29, 47, *157,* 270
Florence, Duomo, 7, 100, 170, 171, *175,* 182,
    198, 205
    Foundling Hospital, 170
    S. Croce, 131, 172
    Uffizi Gallery, 202, *202*
Fo-Kuang, Temple of, Shansi Province, 115
Fontana, Domenico, 21, *191,* 192, 193, 207
Foundling Hospital, Florence, 170
Fox, Francis, 110
Frankl, P., 135, 136, 137, 139, 140, 150, 272
Franklin, Benjamin, 131
Friday Mosque, Isfahan, *101,* 102, *102*
Frontinus, 12, 56, 81, 82, 124
Fumagalli, E., 150, 272

Galileo Galilei, 14, 187, *188,* 213, 226, 227, 228,
    229, 273, 274
Gardette, C.M. de la, *266*
Garisenda Tower, Bologna, 110, *111,* 271
Gauthey, Emiland Marie, 225
Gerona Cathedral, 131
Ghiberti, Lorenzo, 176
Ghini, Giovanni, 172
Giedon, S., 36, 270, 275
Gizeh, Pyramid of Khufu, *29,* 131
    Temple of Khafra, 36, *40*
Glastonbury Abbey, Somerset, *121*
Godwin, George, 239, *240,* 274
Goethe, Johann Wolfgang von, 262
Golden House of Nero, Rome, 57, 63, 70
Gol Gomuz, Bijapur, 178, 202
Gordon, Lewis, 247
Great Pyramid, Gizeh, *29,* 131
    Mexico City, 7, 27
Gregory, David, 179, 182, 195, 198, 200
Grimthorpe, Lord, *see* Denison, Edmund Beckett
Guarini, G., 15, *218,* 263, 269
Guericke, Otto von, 213
Gutenberg, Johann, 171

Haegermann, G., 256, 271
Halikarnassos, Mausoleum of, 27
Hamilton, S.B., 51, 270
Hammurabi, 27
Hampton Court Palace, London, 120, 125, 206
Hanging Gardens of Babylon, 27
Haphaestos, Temple of, Athens, *48*
Herculaneum, Italy, 3, 7
Herland, Hugh, 113, 116
Hero of Alexandria, 14, 16, *17, 18, 20, 21, 22, 83*
Herodes Atticus, Theater of, Athens, *37,* 88, *88*
Herod the Great, King of Judaea, 67
Heyman, Jacques, 46, 47, 50, 115, 150, 156, *157,*
    159, 160, *161,* 163, 164, 198, *200,* 201,
    215, 229, 270, 271, 272, 273
Hindu architecture, 38, *42*
Hipparchus, 14
Hire, Philippe de la, 195, *195,* 225
Holy Roman Empire, 96
Homer, 263
Honnecourt, *see* Villard de Honnecourt
Hontañon, Rodrigo Gil de, 140
Hooke, Robert, 179, 226, *227, 228,* 244
Hospital, Foundling, Florence, 170
    Royal Derbyshire (Infirmary), 267
    Westminster, 267
Hôtel des Invalides, Paris, 225
Houses of Parliament, London, 267
Howard, Castle, Yorkshire, 263
Hurley, William, 154
Hurstmanceaux Castle, Sussex, 120

Igreja do Carma, Lisbon, *130*
Iktinos, 44
Imperial Theater, Vienna, 212
Inca, *see* Peru
India, architecture, 38, *42, 104*
Indonesia, temple in, *30*
Ise Shrine, Japan, 117, 271
Isfahan, Friday Mosque, *101,* 102, *102*
    Sheikh Lutfullah Mosque, *103*
Isodorus, 97, 98
Istanbul, 102
    Blue (Sultan Ahmet) Mosque, 7, *100*
    Suleimaniyeh Mosque, 102
    see also Constantinople

Jacquier, F., 198, 222
Japan, Ise Shrine, 117, 271
Jerusalem, Dome of the Rock, 103, *105, 106,* 129
    Herodean, Temple of, *67*
    Old City, 144
Johnson, Isaac Charles, 261
Jones, Inigo, 178, 206
Julius Caesar, 12, 33, 53, *54*
Justinian, Palace of, Constantinople, 7, 100

Kallicrates, 44
Kancheepuram, temple in, *42*
Karnak, Temple of Amon, 36, 38, *41*
Kayseri, mosque in, *145*
Khafra, Temple of, Gizeh, 36, *40*
Khorsabad, Palace of Sargon, 84
Khufu, Pyramid of, Gizeh, *29,* 131
Kimon of Athens, 44
Kings College Chapel, Cambridge, 161, 163, 272
Kircher, Athanasius, 213, *214*
Knossos, Palace of Minos, 81, 84
Kolossos of Rhodes, 27, 51, *51*
Kyoto, temples in, *118, 119*

La Garisenda, Bologna, 110, *111,* 271
L'Asinelli, Bologna, 110, *111, 112,* 271
Leaning Tower, Pisa, 110
Le-Duc, *see* Viollet-le-Duc, E.E.
Leipzig, Thomaskirche, 213
Leonardo da Vinci, 171, 176, 183, 184, *185, 186,*
    187, 215, *216,* 227
Le Seur, T., 198, 222
Library of Alexandria, 107
  in Constantinople, 107
Lichfield Cathedral, 152, *157,* 160
Lighthouse, Eddystone, 256, *257, 258,* 275
Lisbon, Igreja do Carma, *130*
Liverpool, population of, 237
London, 124, 207, 237
    Banqueting Hall, Whitehall, 206
    Bridge, 95, 247
    Buckingham Palace, 264

Crystal Palace, 143, 150
drainage canal, 262
Great fire of, 70, 178, 205, 208, 226
Hampton Court Palace, 120, 125, 206
Houses of Parliament, 267
St. Paul's Cathedral, 147, 178, *180, 181,* 213
sewers, 208, 262
Westminster Abbey, 131, 141, 159, 161, 264
Westminster Hospital, 267
Westminster Palace, 124, 125
    Great Hall, 113, 115, *116*
Longleat House, Wiltshire, *212*
Louis, Victor, 248
Louvre, Paris, 213
Lübeck, Marienkirche, 120
Lucretius, 171
Lyceum of Aristotle, 13

Macauley, Thomas, 209
Machu Picchu, Peru, 3, 33
Maderna, Carlo, 192
Mainstone, R.J., 101, 176, 270, 271, 272, 273
Malta, Mosta Church, 178
Manchester, population of, 237, 238
Marcellus, Theater of, Rome, 88
Marienkirche, Lübeck, 120
Mariotte, Edmé, 228, 229, 231
Mark, Robert, 150, 159, 272
Maryon, Herbert, 51
Mausoleum of Halikarnassos, 27
Maxentius, Basilica of, Rome, *63,* 129, 131
Melzi, Francesco, 183
Mersenne, Marin, 213
Mexico City, Great Pyramid, 7, 27
Michaelangelo Buonarotti, 192, 193
Michaelis, Wilhelm, 186, 261
Mignot, Jean (Giovanni Mignoto), 137, 139
Milan, Basilica of S. Ambroglio, 120
    Cathedral, 137, *138,* 170
    S. Carlo, 178
Minos, Palace of, Knossos, 81, 84
Monge, Gaspard, 222
Monte Cassino, Monastery of, 129
Morgan, Morris Hickey, 12, 256, 269
Moseley, H., 147, 272
Mosque, Blue (Sultan Ahmet), Istanbul, 7, *100*
    Friday, Isfahan, *101,* 102, *102*
    in Kayseri, Turkey, *145*
    Sheikh Lutfullah, Isfahan, *103*
    Suleimaniyeh, Istanbul, 102
Mosta Church, Malta, 178
Museum of Alexandria, 13, 24
Mussche, H.F., 47, 270
Musschenbroek, Petrus van, 222, 223, 224, *224,*
    235, 236
Mycaene, gateway, *34*
  vault, *35*

Nash, John, 249, 263
Navier, Louis Marie Henry, 147, 228
Needham, J., 269
Nelli, G.B., *173*
Nemorarius, Jordanus, 184, 187
Neri de Fioravante, 171
Nero, Roman emperor, 70. *See also* Golden
    House of Nero
Nestorians, 107
Newcomen, Thomas, 241
New Orleans, 248
Newton, Isaac, 198, 223
New York, Singer Building, 29
Nîmes, Pont-du-Gard, 84
Norwich Cathedral, 142
Notre Dame, Paris, 160

Obelisk, incomplete in Aswan quarry, 38
   Piazza S. Pietro, 21, 189, *191*, 192
   Rameses II, 21
Olimpico Theater, Vicenza, 210, 211
Olympian Zeus, Temple of, Athens, *46, 47*
Orvieto Cathedral, 139
Oxford, Sheldonian Theatre, 178

Paine, Thomas, 246
Palace, Bishop's, Wells, *109*
   Buckingham, London, 264
   Ctesiphon, Iraq, 71, 104
   Diocletian, Spalatum, 72, 129
   Hampton Court, London, 120
   Justinian, Constantinople, 7, 100
   Louvre, Paris, 213
   Minos, Knossos, 81, 84
   Nero, Golden House, Rome, 57, 63, 70
   Sargon II, Khorsabad, 84
   Versailles, 207, 221, 228
   Westminster, London, 113, 115, *116*, 124, 125
Palladio, Andrea, 53, *54*, 178, 202, *203*, 204, 210,
    215, 270
Palladius, 11, 56, 84
Pantheon, Rome, 6, 7, 57, *57, 58*, 59, 63, 71, *72*,
    74, 100, 101, 131, 176, 178, 193, 202
Panthéon, Paris, *see* St. Geneviève
Pappus, 21
Parent, A., 229, 231
Paris, 124, 125, 207
   Hôtel des Invalides, 225
   Louvre, 213
   Notre Dame Cathedral, 160
   Obelisk, 21
   Panthéon, 225
   Pont de la Concorde, 221
   Pont de Neuilly, 221
   St. Denis, Abbey of, 129, 135, 137, 160
   St. Geneviève, *225*

sewers, 208
   Théâtre-Français, 247
Parker, James, 259
Parma, Farnese Theater, *211*
Parsons, W.B., 174, 176, *177, 185, 186*, 273
Partheon, Athens, 7, 15, 44, *45*
Pasley, W.C., 261
Paxton, Joseph, 150
Pericles, 44
Perronet, Jean, 221, 225
Persia, architecture, *101, 102, 103, 272*
Peru, 3, *32, 33*, 64, 141
Peruzzi, Baldassare, 192
Peterborough Cathedral, 142
Pevsner, N., 274
Pharos of Alexandria, 27
Philon of Byzantium, 51
Pippard, A.J.S., 272
Pisa, Baptistry, 182
   Leaning Tower, 110
Plato, 13, 24
Pliny, 11, 12, 51, 56, 68, 263
Plombières-les-Bains, France, 207
Plommer, H., 84, 270
Poggio, Gian Francisco, 170
Poleni, Giovanni, 195, 198, 200, 223
Pompeii, Italy, 3, 7, 88, 122
Pons Aelius, Rome, 95
   Cestius, Rome, 72, 95
   Fabricius, Rome, 95
Ponte de la Concorde, Paris, 221
   de Neuilly, Paris, 221
   -du-Gard, Nîmes, 84
Pope Paul V, 84
   Sixtus V, 84, 207
   Urban VIII, 187
Porta, Giacomo della, 192, 193
Porta Nigra, Trier, 270
Prager, F., 273
Pritchard, T.F., *245, 246*
Procopius, 97, 271
Propylaea, Athens, 45
Ptolemy, 11, 14, 108
Puteoli, 55
Pyramid, Bent, Egypt, 36
   Khufu, Egypt, *29*, 131
   Mexico City, 7, 27
Pythagoras, 13

Quetzalcoatl, Temple of, Teotihuacan, *31*

Ramalli, Agostino, *190*
Rameau, Jean-Phillippe, 267
Rameses II, Obelisk, Paris, 21
   Temple of, Abu Simbel, *39*
    Thebes, *38*
Ramesseum, Thebes, *38*

GENERAL INDEX

Entries in italics are foreign words, names of institutions or titles of books
Page numbers in italics refer to illustrations

Toilets, 84, 124

Tombs, 3, *35*. *See also* Pyramids, Egyptian

Tower of Beauvais Cathedral, 131, 159
   Chichester Cathedral, 152
   La Garisenda, Bologna, 110, *111*, 271
   l'Asinelli, Bologna, 110, *111*, *112*, 271
   Leaning, Pisa, 110
   Winchester Cathedral, 136

Towers for security, 109, 271

Town Swamps and Social Bridges (by Godwin), 239, *240*, 274

Traité de Mécanique (by de la Hire), 195, *195*

Traité de mouvement des eaux (by Mariotte), 229

Transport, medieval, 121, 135, 137, 141
   18th century, 241
   railway, 241
   Roman, 95, 192

Della Trasportatione dell'obelisco Vaticano (by Fontana), *191*, 192

Trasura, 141

Treatise on Strength of Materials (by Barlow), 223

Triangle of forces, *see* Parallelogram of forces

True arch, 36, *37*, 70

Trusses, 114, *115*, 172, 179, *202*, 280

Universities, 108

Vaulted ceilings, lath and plaster, 54, 263

Vaulting, 34, 35
   construction, *38*, *143*, *146*
   fan, 133, 162, *162*, 272, 273

   forces in, 160, *161*
   Gothic, 129, 131, 140, 142, *146*, 160, *161*
   Middle Eastern, *102*, *103*, *144*, *145*
   quadripartite, 133
   Renaissance, 172, 193
   Roman, *63*, *64*, *65*, 70, 71

Veneer, 263

Ventilation, 90, 267

Vigiles, 70

Vitruvius Brittanicus (by Campbell), 170

Voyages of discovery, 171

Waste disposal, *see* Sewage disposal

Water supply, medieval, 124
   18th century, 239, 265, *266*
   pre-Roman, 81
   Renaissance, 207
   Roman, 53, 81, *82*, 96, 207
   see also Aqueducts

Wattle and daub, 114

Weights, *see* Hoisting of weights

Whispering galleries, 213

Windmills, 108

Windows, 122, 206, 209, *211*, *212*

Wonders of the World, The Seven, 27, 51

Wood, *see* Timber properties

Wren Society, 179, 273

Young's modulus, 230, 281

Ziggurats, 27